SHIPBUILDING, NAVIGATION AND THE PORTUGUESE IN PRE-MODERN INDIA

Shipbuilding, Navigation and the Portuguese in Pre-modern India

K.S. MATHEW

Routledge
Taylor & Francis Group
LONDON AND NEW YORK

First published 2018
by Routledge
2 Park Square, Milton Park, Abingdon, Oxon OX14 4RN

and by Routledge
711 Third Avenue, New York, NY 10017

Routledge is an imprint of the Taylor & Francis Group, an informa business

British Library Cataloguing in Publication Data
A catalogue record for this book is available from the British Library

Library of Congress Cataloging in Publication Data
A catalog record for this book has been requested

ISBN: 978-1-138-09476-5 (hbk)
ISBN: 978-1-315-10118-7 (ebk)

Typeset in Dante MT Std 11/13
by Ravi Shanker Delhi 110 055

MANOHAR

Contents

Acknowledgements

The idea of working on the different aspects of naval architecture and navigation in pre-modern India, especially during the sixteenth century when there was close interaction between the Indians and the Portuguese, took root in my mind ever since I started research on Indo-Portuguese history in the 1970s. This was envisaged to make a comparative study of the indigenous and Portuguese techniques of shipbuilding and navigation and to identify the mutual give and take if any. This long cherished plan was materialized easily with the award of a Short-term Research Fellowship by the Fundação Oriente, Lisbon and Emeritus Fellowship by the University Grants Commission, government of India. My heart-felt thanks are due to the president and the director of Fundação Oriente and to the University Grants Commission for providing financial assistance to complete my researches.

I express my sincere thanks to all those who extended a helping hand in my attempts to collect relevant sources of information from the various repositories of documents in Europe and India: Archivo Nacional da Torre do Tombo, Bibliotheca de Marinha, Bibliotheca Nacional, Lisbon, Leiden University Library, The Netherlands, Goa Historical Archives, Panaji, and Tamil Nadu State Archives, Chennai. I place on record my indebtedness to the authorities of these institutions for their goodwill.

I am greatly beholden to the Manohar under the stewardship of Mr. Ramesh Jain for bringing out the book neatly.

Kurumulloor K.S. MATHEW
21 September 2016

CHAPTER I

Introduction

The Relevance of the Study

Scholarship on the maritime history of India has covered a number of aspects such as imports and exports of the sea-borne trade; merchant groups and guilds; activities of the merchant companies such as the English East India Company, the United Dutch East India Company, the Danish East India Company, and the French East India Company; merchant financiers, brokers and middlemen; port–hinterland relations; the dynamics of the relation between the colonies and metropolis; weight and measures; and Indo-European coinage. The Indian Ocean figures as the chief thoroughfare for the international sea-borne trade during the pre-modern period. Some studies have been conducted also on migrations of culture, peoples and even agricultural products across the Ocean. In the wake of the discovery of the vast riches of India, especially peninsular India, the rulers and merchant companies of Western Europe vied with each other to gain a foothold in India at the risk of the loss of life of hundreds of mariners on the high seas in their internecine naval confrontations. The role played by port towns and hinterlands has also become an interesting subject. Port towns in the colonial period turned out to be centres of extraction of the surplus produce of the hinterland and places of storage of foreign items of exchange besides offering space for bureaucracies and administrative offices. Life and culture of the coastal and island society with their eyes fixed on the sea constitute another aspect of maritime history that can attract the attention of the social historians as well as scholars deeply interested in psycho-history. The Indian Ocean instead of separating peoples, could also bring various peoples together to pave the way for the emergence of a cosmopolitan society. No interactions or confrontations across the ocean could be effected without the various types of seacraft mariners who steered them with the available technology at their disposal.

Scholars of maritime history committed to highlight the maritime heritage of India in terms of science and technology are now busy collecting data on the materials for the construction of ships, instruments of navigation and the technology applied in these areas. This will help research in the history of science and technology to widen its scope by entering into maritime space and activities. The space which we occupy or which our people in the past occupied becomes enlightened when we make comparative studies. Therefore, scholars have been showing interest in comparing the science and technology of Indian shipbuilding and navigation with those of Europe. As the Portuguese were the pioneers who crossed from the Atlantic to the Indian Ocean via the Cape of Good Hope they had early interaction with Indian shipwrights and mariners, in the sixteenth century. The present study takes into account Indian shipbuilding and navigation and Portuguese maritime activities in India in the sixteenth century in a comparative approach. Interactions took place mostly in the littoral of peninsular India with a view to extracting maximum surplus produce of spices and black gold at the cheapest rate possible to maximize the profit under the ideology of mercantilism. Hence the study is addressed chiefly to the seafront of peninsular India with special emphasis on the Malabar coast.

A serious study of the history of shipbuilding and navigation in India before she came in to contact with the Europeans especially the Portuguese will definitely go a long way in making ourselves conscious of our heritage and its worth in general. We will be able to understand the importance of the infrastructural assets. Teak wood which had the potential to replace the oak of Europe for the construction of ocean-going vessels, was one of the important materials used for shipbuilding in pre-Portuguese and Portuguese India. Its worth was understood not only by the Portuguese but also by other European powers that came to trade with coastal India. The English prohibited the felling of this tree and made it a monopoly of the government of British India. This prohibition was in force even at the close of the twentieth century in Kerala. The qualities of this timber made Portuguese experts write that India was the most congenial place for building sea-going vessels. Our seamen used to visit the East African coast, the Red Sea and Persian Gulf regions. If the ships that were built in India before the sixteenth century did not possess the qualities to withstand the vagaries of the high seas, it would have been impossible to conduct such voyages. Vessels built in India were of high quality and could cross the high seas. This speaks for the expertise of the shipwrights and high qualities of timber used as well as the technology mastered by the people of the past.

In a period when no sophisticated instruments were known, Indian vessels used to visit distant destinations with commodities. One could look into their techniques of haven-finding and their instruments of navigation. Similarly various other aspects of navigation during the period prior to the sixteenth and during the sixteenth century can usefully be studied to highlight the maritime heritage of India. This will enrich our knowledge of the past and help us enquire whether or not indigenous techniques of navigation were retarded by the influx of European methods of navigation.

Comparative study will help us assess the stage of development reached by Indian mariners when the Europeans began to interact with our technology. One could also identify the innovations, if any, made by Indian seamen due to this interaction and vice-versa. Very little work has been done in this direction though isolated studies either on Indian shipbuilding and navigation or Portuguese shipbuilding and navigation have already appeared.

Further, we will be able to collect more data related to the subject and provide material for further research in the field of the history of science and technology. Therefore it is judged relevant to take up the study of *Portuguese and the Shipbuilding, Navigation in Pre-modern India*. It has been said that the life of the people on board a ship was distanced from death only by the thickness of the timber of the ship.

Radha Kumud Mookerji published his *Indian Shipping: A History of the Sea-borne Trade and Maritime Activity of the Indians from the Earliest Times*, using Sanskrit, Pali, Tamil, Bengali and Persian sources. He also used translations of Chinese and Japanese material for the preparation of his Ph.D. thesis submitted to the Calcutta University.

The thesis was later developed and was brought out in the form a book in 1910. He divided the period of his study into Hindu, and Mahomedan periods and added a small chapter on 'Later Times'. While dealing with sea-borne trade he makes mention of indigenous techniques of shipbuilding.

B. Arunachalam has done considerable and commendable work on indigenous systems of navigation. His publications are important for the study of Indian navigation and cartography. He has dealt with the haven-finding art practised by the Indian navigators. His publications include among others the following: (1) *Heritage of Indian Sea Navigation* (2) *Essays in Maritime History*, in two volumes (3) *Chola Navigation Package* (4) *Navigational Hazards, Landmarks and Early Charting: Special Study of Konkan and South Gujarat* (5) 'Chola Navigation Package: Indigenous Traditions of Indian Navigation – Report of the CSIR Project', (6) 'The Haven – Finding Art in Indian Navigational Traditions and Cartography' (7) 'The Chola Mode of Navigation in the Northern Indian Ocean,' (8) 'Traditional Sea and Sky wisdom of Indian Seamen and their Practical Applications', (9) 'Traditions and Problems of Indian Nautical Cartography' and (10) 'Timber Traditions in Indian Boat Technology'. He collected a number of oral traditions from various parts of coastal India and the adjacent islands. He laid hands on available manuscripts related to indigenous techniques of navigation.

K.M. Mathew worked on the history of the Portuguese Navigation (1497-1600) and published *History of the Portuguese Navigation in India* (1988), on the nautical knowledge of the Portuguese, their cartography and naval battles. The book deals with coastal forts, shipwrecks, the organization of the navy, the decline of the Portuguese naval policy, shipwrecks, and so on.

Xavier Mariona Martins worked on Portuguese shipping and shipbuilding in Goa from 1510 to 1780 for a doctoral dissentation at the Goa University. He lays emphasis on shipbuilding in Goa and aspects like the naval dockyard, various types of ships, the mechanism of Portuguese shipping and so on.

Abdul Khader for the M.Phil, under the guidance of the present author worked on indigenous technique of shipbuilding chiefly taking into account the oral tradition current on Malabar Coast.

B.K. Apte worked on the *History of the Maratha Navy and Merchant-*

ships (1973). This work is helpful to make a comparative study of naval and merchant ships.

K.S. Mathew edited the proceedings of a seminar held in 1997 on shipbuilding and navigation in the Indian Ocean regions. A few articles in this work shed light on Indigenous shipbuilding and navigation. Similarly another book edited by the same author on maritime studies contains an article on the haven-finding art and another on shipbuilding in India. In fact a comparative study of the techniques of shipbuilding and navigation for the period of the sixteenth century based on contemporary sources in Indian and European languages is a desideratum which will highlight the Indian heritage and throw light on the give and take.

A few sources in epigraphy and numismatics help us reconstruct the history of indigenous shipping. Representations of ships and boats found in cave no 2 at Ajanta dated between AD 525 and 650 give indications about the mast and the shape of ocean going vessels. Similarly the coins issued by Pallavas and Ishwakus also furnish us with some glimpses of Indian ships.

India has not developed marine archaeology and underwater excavation on par with the achievements of scholars of Fremantle near Perth in Australia where ancient ships are recovered from the bottom of the sea, rebuilt and preserved in museums. Of course some work is being done by the Marine Archaeology centre of the National Institute of Oceanography where a few scholars, expert in diving, have collected some parts of a Portuguese ship wrecked in Goan waters around 1651. Some attempts have been made in Poompuhar to unearth the old town of Poompuhar. But naval archaeology in India has yet to develop.

Among written sources *Yuktikalpataru*, a Sanskrit tract by Bhoja Narapati of the early medieval period of India is the earliest treatise on shipbuilding. It contains many details about the various sizes of ships and the materials for the construction. Though this is of an early period as far as the subject of our treatment is concerned, it can be consulted judiciously for the study on indigenous shipbuilding.

There are a few near contemporary sources in Tamil. *Navoi Sattiram (Navái Sastram)* in the McKenzie collection, though primarily astrological in character, has some details regarding boat building and the felling of trees at the right time. This was written by Trikuta Nambi. There is a host of information related to the quality of timber for ship/boat building. A tract edited and published by S. Soundrapandian. *Kappal Sattiram* of Tharangampadi (1620) and critically edited by T.P. Palaniyappa Pillai was probably of a century and a half after the *Navoi Sattiram*. It deals with timber for the building of ships. The Tamil ballad, *Kulathurayyan Kappalpattu*, a late eighteenth-century treatise, deals with the search for suitable timber in the southern Travancore forests. A Tamil work *Calavettu-Pattu* of Nagapattinam, Nagore classifies timbers as masculine, feminine and neuter as discussed in the *Yuktikalpataru*.

No serious sources in Malayalam related to shipbuilding have so far come to light. A ritualistic song connected with the worship of Kannaki in some Sreekurumba shrines of Malabar provides us with some details on shipbuilding in the medieval period. The song is known as *Kannakiyum Cheermakkavum* published by C.M.S. Chandera in 1973. It refers to the details of ship and the launching thereof.

Ars Nautica written in Latin, in 1570 by Ferdinandus Oliverius (Fernando Oliveira) is one of the most important documents on shipbuilding and navigation. This manuscript containing 236 folios (472 pages) is preserved in the Library of the University of Leiden, under the title *Ex Biblioteca Viri Illustris Isaaci Vossi*, no. 15. It gives details of techniques of shipbuilding and navigation in Portugal. Some reference is found also for the Portuguese activities of shipbuilding and navigation in the Indian Ocean regions. It has three parts.[1] The existence of this manuscript was brought to the notice of the public by Luis de Matos.[2] The preface of the work *Ars Nautica* deals with the advantages of navigation. The author busies himself with cartography, instruments of navigation, demarcation and calculations of route besides meteorology in the first part of the work. The details of shipbuilding are discussed in the second part of the manuscript. This contains the earliest scientific discussion on shipbuilding till date.[3] He writes also about the history of navigation based on works of Roman and Greek authors. He concentrates, in the third part, chiefly on the physical and moral qualities of the mariners, their discipline and their diet on board.[4]

This is the earliest surviving written document on naval

architecture. Fernando Oliveira was born in 1507 in Aveiro in Portugal, and began his education under the Dominicans. At thirteen, he joined the Convent in Évora where he attended the classes of André Rezende. The strict discipline of the convent and his eagerness for knowledge made him run away from the convent when he was twenty-five years old. Then he went to Castella. Later, he came to Lisbon and wrote *Grammatica de Lingua Portuguesa* in 1536. He was in touch with João de Barros, the humanist and author of the *Decadas da Àsia*. He left for Italy in 1540 or 1541. He returned to Portugal in 1543.Then he visited France and England. While in France, he got interested in maritime activities. He must have been in England in 1546. He witnessed the corruption in Rome and saw the iconoclastic attitude of the Reformation in England under Henry VIII. He gradually developed an aversion towards all religious orders especially the Dominicans.

He reached Lisbon in 1547 carrying a letter from King Edward VI of England to John III of Portugal. There he renewed his contacts with André Rezende whom he had met in Évora. The inquisition was officially established in 1547. Fernando Oliveira who had run away from the Dominican Order was brought to the Inquisitor, was judged in 1548 and was imprisoned by the Inquisition. At last in 1550, he was transferred to the monastery of Jeronimo. He was allowed to practise as a priest, and put on clerical dress and tonsure by the intervention of Cardinal Infant Henrique.[5]

The systematic and theoretical explanations with general norms make his work scientific and unique. Some of the norms of shipbuilding followed in Portugal are compared with those of other countries since the author had travelled a lot and saw ship-building methods in various parts of Europe. He was also familiar with the literature available in Latin.[6] He was able to correlate the design of a vessel with hydrodynamics and other principles of motion. While dealing with vessels propelled by oars, he emphasizes the relation between the hull of the vessel and its movements in water.

The Collection of *Documenta Indica* edited by Georg Schurhammer and Vicki in 18 volumes (1948-88) contains reports sent by the Jesuit missionaries traveling on board the Portuguese ship. These letters are quite important to deal with the medical care and other aspects of the life of the people on board the Portuguese ships. The letters are chiefly in Latin, Spanish and Italian languages.

Francisco Pyrard de Laval who spent a few years in the Indian Ocean regions wrote a narrative of his journey to India between 1601 and 1611. This book in French (third French edition was released in 1619) has been translated into Portuguese and annotated by Joaquim Heliodoro da Cunha Rivara. It was published in 1944 in two volumes. The English translation was brought out by Albert Gray in three parts under the auspices of the Hakluyt Society, London. An Indian reprint was published by Asian Educational Services in 2000. Chapter XIV in vol. 2 of this book has a description of Portuguese ships and the life of the crew on board, which is of great importance for this study.

John Huyghen van Linschoten was in India from 1583 to 1589 and reached Lisbon in January 1592. His *Itinerario* was allowed to be published in 1596 by the States-General. He gives details of the Portuguese shipping in his book, written originally in Dutch, and later translated into English in 1598. Hakluyt Society brought out the book in two volumes in 1885 and an Indian reprint was made available in 1988 by Asian Educational Services, New Delhi.

Ludovico di Varthema of Bologna, Italy was on the Malabar coast from 1502 to 1508. He wrote an Itinerary of his voyage to India in which he makes mention of sea-going vessels of Malabar. The original work in Italian was published in 1510. John Winter Jones brought out an English translation of the same in 1863 for the Hakluyt Society. Richard Carnac Temple added a discourse on Varthema and published it in 1928. An Indian reprint was published by Asian Educational Services in 1977.

We have the log book of the first voyage of Vasco da Gama and the anonymous narrative of the voyage of Pedro Álvares Cabral which too shed some light on Portuguese shipping in general. The sixteenth-century chronicles of Barros, Diogo do Couto, Gaspar Correa, and Fernão Lopes de Castanheda contain a lot of information on Portuguese shipping. Besides these general works, there are a few specific sources related to naval archaeology.

Extant documents related to Portuguese shipbuilding, preserved in various repositories in Europe, throw invaluable light on the history of naval architecture. They become essential in the absence of shipwrecks or models of ships.

Establishments for careening ships were started by the Portuguese in Cochin right at the beginning of their contacts with the king of Cochin in whom they found a well-disposed ruler. But installations for building ships seem to have come up only by the end of the first decade of the sixteenth century. There is clear evidence of the building of the ship *Sta. Catarina do Monte Sinai* in the Portuguese shipyard in Cochin between 1511 and 1513. This vessel was of 800 tons.[7] As soon as Goa was conquered by Afonso de Albuquerque arrangements were made to build ships in Goa.

An important work on naval architecture, written in the third quarter of the sixteenth century, by Fernando Oliveira is O *Livro da Fabrica das Náos*. He makes mention of his Latin work, *Ars Nautica* in this book.[8] He explains the antiquity of shipbuilding in the introductory chapter and, in the second chapter deals with the types of timber suitable for the building of ships, in Europe and abroad. He suggests that local carpenters should be consulted in selecting appropriate timber. He writes that strong woods should be used for frames and flexible ones for planking, solid and close-grained below the water, light in the upper parts and yards: long, straight and clean for masts.[9] He praises the qualities of the wood found in India- *angeli* (*Angelim arabora*) and teak (*Tectona grandis* or *Andira racemosa*) as extremely useful for the construction of ships.[10] He explains why freshly cut wood should not be used for the construction of ships.

According to Oliveira, iron nails should be used, since they are cheaper, longer-lasting, and stronger than wooden nails. Wooden nails need more boring of the planks in the construction of vessels and this weakens the planks. Larger vessels, with thicker planks, require long-lasting and strong fastenings which wooden nails cannot provide and therefore the Portuguese ships carrying a lot of cargo from India to the West required iron nails. Copper nails could be used, but they are costlier than iron nails.[11]

The author speaks of caulking with oakum in order to fill the joints in planking, and to close and fill-in the seams of the ship. The oakum used by the Portuguese in caulking was soft. It could be compacted. It swelled when it was wet, it accepted pitch or grease or any other sealant to stop water. Pitch, which resists water and preserves oakum was added after the ship was caulked. Fernando Oliveira further explains how good pitch was made.[12] He describes how Indians made the filler, called *çaragaste* for ships from a certain powdered earth. Another sealant, called *galgata* paste, was also used in India to treat the seams of the ship which was made with virgin lime and oakum kneaded with olive oil. This sealant served mainly as protection against shipworms.[13] He deals with vessels of various types in Chapter V. It is generally agreed that the tonnage of vessels in the Indian Ocean regions increased with the increase of trade conducted by the Portuguese. Probably being aware of the on-going controversy about vessels of greater tonnage, Fernando Oliveira defends the need for constructing larger vessels for a greater volume of commodities. Vessels of a large size were less liable to sink than narrow ships. For example, if a plank is thrown flat upon water, it will remain afloat. If the same plank is thrown into water on its edge, it will immediately sink until it attains the normal position. Even though the weight of the plank is the same, the different reactions are on account of the resistance of the water beneath. A wide ship has more water below it than a narrow ship and hence the resistance of water will keep the vessel afloat. The same point is clarified from another angle. Water beneath a wide vessel or plank weighs more than the ship or plank. Air being lighter than water, the wide-bodied ships contain a large volume of air. So, water under big ships will sustain the ships floating and keep them away from sinking. He gives further reasons why the ships of the India run (*Carreira da India*) should be of greater tonnage.[14] He furnishes practical designs of ships of various types.

There is an interesting work under the title *Coriosidades de Gonçallo de Sousa (1580-1600)*. The manuscript copy of this work is preserved in the Biblioteca da Universidade de Coimbra. It contains a lot of information on naval matters related to the construction of ships of the India run and also of oriental ships.

Another practical work of the early part of the seventeenth century is *O Livro Primeiro da Architectura Naval* (*c.*1600-25, Mss. Biblioteca da Academia de História, Madrid) by João Baptista Lavanha of Jewish ancestry who was born in the middle of the sixteenth century in Lisbon. When the Portuguese Crown was brought under Spanish kings, he served Dom Philip II, III and IV of Spain. He taught these monarchs mathematics and was then appointed on 25 December

1582 to teach cosmography, geography, topography and mathematics at the royal court. He took great interest in mathematics and joined the newly established Academy of Mathematics of Madrid. He was appointed Chief Cosmographer from 13 February 1591 in Lisbon but returned to Madrid in 1599 where he died in 1624. He wrote the *Regimento Nautico in* 1595, the *Naufragio da Nau S. Alberto* in 1597 and the *Tabuas da largura ortiva do sol in* 1600.

The only book on naval architecture written by Baptista Lavanha is *O Livro Primeiro da Architectura Naval,* composed between 1607 and 1616. The work has been brought out in facsimile by Academia de Marinha, Lisboa, in 1996. After dealing with architecture in general, he speaks of naval architecture in the fourth chapter. He devotes another chapter to the materials used in shipbuilding and discusses the seasons in which trees should be felled. Other materials used in shipbuilding are discussed in the subsequent chapter. The design of the keel, stem, sternpost, master frames, keel blocks, the fashioning of keel, drawing of moulds, fashioning of transom and so on are treated in detail in the next part of the treatise.

Lavanha stresses the need for a thorough knowledge of arithmetic, geometry and mechanics for a naval architect, who should also have some notion of astronomy besides being a good draughtsman. Astronomy helps him know the movements of the sun and the moon and the winds which are essential for choosing a suitable place for the shipyard. Knowledge of other languages enables the naval architect to be familiar with the progress of his profession in various parts of the world. Lavanha insisted on making a sketch of the ships to be built and discussing the same with the experts before starting the work. It was probably this approach that made the Portuguese viceroy in India, during the first quarter of the seventeenth century, send a model of the *nau* with three or four decks to Lisbon for discussion. Lavanha was an engineer of greater precision than Fernando Oliveira and so his work excels in theoretical and practical aspects over that of the latter.[15]

Lavanha speaks of the use of teak and *angelim* for shipbuilding in India in place of cork, oak, stone pine and maritime pine of Europe. He enumerates the necessary qualities of timber to be used in the building of a ship. Timber should be tough, dry, of bitter and resinous sap and pliable. The impetus of the sea and winds can be resisted by tough timber. Dry timber does not rot with the dampness of the waters. If the sap of timber is resinous, it can get rid of water and the bitterness of the sap keeps shipworms away. Pliability is a quality much needed for bending the planks in the construction and avoiding a split. All these qualities, not always found in any single tree, all occur in *angelim* and teak grown in Malabar. The timber of these trees is incorruptible and it appears that nature created them exclusively for shipbuilding, comments Lavanha.[16]

Some external and general signs to understand the nature of timber are given by Lavanha. The fruits and leaves are to be checked. If a tree has rough bark, crisp leaves and hard fruit, the timber of that tree would be strong and dense. Trees with a smooth bark, soft leaves and fruit, provide soft and porous timber. Those trees that grow slowly provide stronger timber than those that grow fast. Similarly, trees with resinous and bitter sap are capable of resisting water and shipworms. Those with a long life rot slowly. Architects are advised to identify suitable timber from areas outside Portugal based on these general principles and to use them for different parts of ships: the skeleton, planking, and so on.

Lavanha gives details of the season during which trees could be felled for shipbuilding. He emphasizes these aspects because the decay of the timber of a house on land is of less importance than the decay of a plank in a vessel moving in water. He suggests that trees be cut after they have given their fruits and have gathered sufficient strength to yield fruits again. Considerations of the sun and the moon are to be given due importance in choosing the appropriate season for felling trees. Summer is preferred for felling trees for building vessels but, according to Lavanha, the phase of the waning moon is to be chosen.

There are a few more near contemporary sources (seventeenth century) which too shed a lot of light on shipbuilding and navigation. They are:

a. *O Livro de Traças de Carpintaria*, written in 1616 by Manuel Fernandez, a well-known writer on naval architecture. The manuscript is in the Bibliotheca de Ajuda, Lisboa. It furnishes designs for ships with scientific measurements. A part of it was published by Estanislaus de Barros but without any technical study in 1933. It has been now brought out in its entirety with facsimile by the Academia de Marinha, Lisboa.

b. *O Livro Nautico ou meio pratico da Construção dos Navios e Gales antigas, e memorial de varias cousas*, is a manuscript preserved in the Biblioteca Nacional de Lisboa. It contains precious details about the construction of ships of 600 tons, galleys of 500 tons, caravels and other types of vessels.

c. *As Advertencias de Naveguantes (1625-1640).* The manuscript is preserved in the Biblioteca de Cadaval de Marcos de Aguiar. It contains some important details on the equipping of ships and their construction and a vocabulary of nautical terms in Portuguese.

d. *O Tratado do que deve saber hum bom Soldado para ser bom capitam de Mar e Guerra.* This manuscript is preserved in the Biblioteca da Universidade de Coimbra. It contains some details on shipbuilding especially of galleys, and nautical terms.

The subject matter of this work will be discussed in the following sequence.

The second chapter deals with shipbuilding and navigation in India prior to the dawn of the sixteenth century. The use of timber will be one of the important aspects that will be covered in this chapter. Different varieties of timber are suggested for different parts of the ship. Curing of the timber from the trees felled as well as the transportation of the same from the interior places to the centres of shipbuilding will enlighten the readers on the means of transportation of timber. Use of nails whether of iron or of timber (wooden pegs) shall also be discussed here. The indigenous technique of caulking which was highly praised by the Portuguese shall be highlighted. A few words about the shipwrights and their expertise will get due attention in this chapter. .The launching of the vessels and other related ceremonies shall also be studied. The role played by human energy, by *Khalasees* will be discussed. Similarly, attempts will be made to study the centres of shipbuilding prior to the arrival of the Portuguese. It will be our endeavour to discuss briefly the different types of ships built on the western coast of India. Further, techniques of navigation adopted by the Indian mariners shall be discussed. The instruments used in the art of haven-finding while the vessel is on the high seas shall be highlighted in this chapter.

With a view to understanding the maritime activities of the Portuguese on the western coast of India and the interactions of Indians with the Portuguese, a short description about the arrival of the Portuguese and their settlements in India will be studied in the third chapter. They entered on friendly relations with a few native rulers and obtained permission to set up their factories and fortresses in strategically important places like Diu, Daman, Bassein, Dabol, Chaul, Goa, Cannanore, Calicut, Cranganore, Cochin and Quilon. They set up shipbuilding centres attached to their establishments on the western coast of India. They looked forward to getting access to the interior places from where various varieties of timber suited to the building of ships could be brought to the centres of shipbuilding. After a long voyage from Lisbon to India they needed installations for careening the ships, and even building ships. They always took into account the availability of suitable timber when they set up the installations for shipbuilding.

The fourth chapter will deal with Portuguese shipbuilding in India giving emphasis to naval architecture. Different centres of shipbuilding under the Portuguese will also be discussed here. Besides, an attempt is also made to highlight various types of vessels built by the Portuguese in India. The tonnage of the ships built by the Portuguese differed greatly in tune with the unprecedented and consistent increase in the volume of commodities exported from India to Portugal. The early vessels especially those under the command of Vasco da Gama for his first voyage were of minimal tonnage. The number and status (quality) of the passengers too increased during this period. The long duration of voyage from Portugal to India and vice-versa necessitated the enhancement of the tonnage of the vessels. More amenities had to be provided in the ship taking into account the nature and number of the passengers. Similarly on account of the naval encounters with the competing west Europeans for a share in the maritime trade of India prompted the Portuguese to mount more cannons and other equipments for naval battles. This had to be reflected in the volume and architecture of the vessels in the sixteenth century. The Portuguese used more Indian timber especially teak, *angeli* and so on which were felled in appropriate seasons according to the suggestions of local carpenters whose services were sought by the Portuguese. The masts were made of timber different from those used for the hull and other external parts.

The fifth chapter deals with Portuguese navigation and their exclusive claim in the Indian Ocean regions. Portuguese navigators used a variety of equipments for voyage from the Atlantic through Indian Ocean or Arabian Sea to reach the western coast of India and on return. Different types of veils and oars were used to harness the natural and human energy for the propulsion of the vessels. In the absence of modern navigational equipments, they had their own equipments to decipher the direction based on the position of stars. They used astronomical techniques for navigation. They had special instruments and techniques for the calculation of time, depth of the sea, etc. Messages from the captain in chief of a fleet had to be passed on to the various captains of the ships constituting the fleet.

The Portuguese interested in maintaining a monopoly in the Indian Ocean regions asserted their exclusive right of passage. The king of Portugal right at the beginning of the sixteenth century assumed a controversial title *Senhor da Navegação e conquista da India*.... Others interested in navigating in the Indian Ocean were constrained to take a pass from the Portuguese. This was extended even to the Indian rulers. So, some discussions on this aspect shall be found in this chapter.

The passengers and crew of the Portuguese vessels coming to India and returning to Portugal constituted a microcosm. They had to be accommodated for several months on board. They had their own hierarchy of authority and values. They contracted some diseases peculiar to the tropics besides the common diseases. The mental, religious and physical aspects of their life for several months should be taken into account. Some entertainment had to be provided besides religious performance to keep the passengers under control. Therefore discussions of these aspects will be done in chapter six entitled 'Life on Board a Portuguese Ship of the Sixteenth Century'.

After seeing the various aspects of Indian shipbuilding and navigation and discussing the various parameters of the Portuguese maritime activities, it will be convenient to make a comparative study delineating the aspects of 'give and take' if any and assessing the Indian maritime heritage on the canvas of science and technology in a historical perspective. This will be done in chapter seven.

A select bibliography which will be helpful for further in-depth studies in the field of science and technology, besides maritime history is provided.

NOTES AND REFERENCES

1. The present writer during his visit to the Leiden University library consulted the work and managed to get the entire manuscript photocopied through the good offices of Mr. Binu John working in the Leiden University for his doctorate in Indo-Dutch history under the TANAP programme. It is an unedited manuscript. There was some announcement from the Academia de Marinha (Lisbon) that they would be taking up the translation from Latin to Portuguese and publish it. The present writer has been elected in 2005 as an associate member of the Academia de Marinha on account of his contribution to the Portuguese maritime activities. To the best of his knowledge, this has not been edited, translated nor published till now.
2. Luís de Matos, 'O Manuscrito Autógrafo da Ars Nautica de Fernando Oliveira', *Bol. Intern. Bibl. Luso-Brasileiro*, 1, no. 2 (1960).
3. Ref. for details, H. Lopes de Mendonça, 'O Padre Fernado Oliveira e a sua Obra nautica' *Memorias da Academia das Ciencias*, Lisboa, 1898, tomo vii, parte ii.
4. João de Gama Pimentel Barata, *A Ars Nautica do Padre Fernando Oliveira: Enciclopedia de Conhecimentos maritimos e primeiro tratado cientifico de construção naval (1570),* Lisboa, 1972 (*separata).*
5. Henrique Lopes de Mendonça, 'O Padre Fernando Oliveira e a sua Obra nautica', *Memorias da Academia das Ciencias*, Lisboa, 1898. He makes a detailed study of Fernando Oliveira in 145 pages.
6. Barata, *A Ars Nautica do Padre Fernando Oliveira, op. cit.*, p. 10.
7. The exquisite finish of the vessel is described by Garcia de Resende re. *Hida da infant dona Beatriz pera Saboya.* This vessel was used by Infanta Dona Beatrix, duchess of Saboia and daughter of D. Manuel I of Portugal.
8. Fernando Oliveira, *O Livro da Fabrica das Naos,* Lisboa, 1991, p. 152. The Manuscript of this work is preserved in the Bibiliotheca Nacional de Lisboa. This must have been composed between 1570 and 1580
9. *Ibid.*, p. 146.
10. *Ibid.*, p. 149.
11. *Ibid.*, p. 150.
12. *Ibid.*, p. 152.

13. *Ibid.*, p. 153. Detailed description of the making of *galagata* paste is given by Adelino de Almeida Calado (ed), *Livro que trata das cousas da India e do Japão*, a sixteenth century manuscript of Nycolão Gonçalves preserved in the Municipal Library of Elvas under no. 5/381, Coimbra, 1957, Calado speaks of *çaramguste* ref. pp. 67-9.
14. Oliveira, *O Livro da Fabrica das Naos*, pp. 162-5.
15. João Bapitsta Lavanha, *Livro Primeiro da Architectura Naval*, Lisboa, 1996, p. 191.
16. *Ibid.*, p. 141.

CHAPTER II

Shipbuilding and Navigation in India Prior to 1500

Malabar in the medieval period was blessed with the best suited timber for the manufacture of ships. The Portuguese writers of the sixteenth century used to say that the best variety of timber for building ships was available in Malabar so much so that one could get the impression that 'nature created them [teak- *Tectona grandis* or *Andira racemosa* and *Angelim amargosa* or *Andira vermifuga*] for naval architecture'.[1] This prompted the Europeans to set up centres of shipbuilding in various parts of coastal Malabar from the beginning of the sixteenth century. Indian mariners too used to visit various remote areas of the Indian Ocean regions like the East African coast, West Asian regions and even Malacca. This presupposes an expertise in deep sea navigation. On his maiden voyage to Calicut Vasco da Gama met Indian mariners at Melinde. It is reasonably argued that a few of them were from the Malabar coast. The present chapter will discuss aspects of shipbuilding, and then navigation, before the arrival of the Portuguese. First we will have a cursory look at the various indigenous sources throwing light on these aspects.

Sanskrit Sources

The *Vedanga Jyotisha* of about the fifth century BC, Sanskrit astronomical texts, generally discuss various kinds of instruments for time measurement and astronomical observation.

Aryabhatta in the fifth century AD described variety of interesting instruments.[2] Brahmagupta in the first quarter of the seventh century devoted an entire chapter of his *Brahmasphuta siddhanta* for discussing instruments like water clocks of the sinking bowl type, several kinds of sundials with dials projected on plane or hemispherical surface, instruments for measuring altitude like the quadrant and the double quadrant, and so on.[3]

Yuktikalpataru,[4] is allegedly from the pen of the polymath Bhoja, who ruled Malwa in the first half of the eleventh century. Scholars have shown that the text was not composed by Bhoja nor has it any value for the history of Indian shipping.[5]

There are two important Sanskrit works dealing with the instruments for astronomical calculation. The first is *Yantraratnavali* composed by Padmanabha in central India in 1423 and the second, called *Yantraprakasha,* was written in 1428 by Ramacandra Vajapey in northern India. *Yantraratnavali* introduces two instruments, one indigenous and the other a variant of the astrolabe. Padmanabha deals with an unusual variety of astrolabe known as the southern astrolabe. He further introduces *Dhruvabhramsa yantra,* a kind of nocturnal, similar to the European nocturnal. It consists of an oblong metal plate with a horizontal slit at the top. A four-armed index is pivoted to the centre of the plate, around which there are concentric circles containing various scales. With the help of this instrument, one can see the Pole Star and Beta Ursae Minoris through the slit. When these two stars are sighted in a straight line by appropriately tilting the instrument, the arms of the index will point respectively to the sidereal time in *ghatis* and *palas*, to the *lagna* or ascendant for this moment, and to the culminating point of ecliptic. Thus with this instrument one could read off from the dial, for any given moment at night, the corresponding sidereal time, ascendant, and culmination. The reverse side of this instrument usually contained a quadrant (*turiya-yantra*) for measuring the solar altitude during the day, or a horary quadrant for measuring time directly.

In *Yantraprakasha* Ramacandra gives a detailed description of a sand clock (*kaca-yantra)* which measures the Indian time unit, the *ghati,* of 24 minutes. European sand clocks are made to measure the hour of 60 minutes, its multiples or fractions. The first four chapters of *Yantraprakasha* speak of the astrolabe. He further introduces the *cudayantra* to identify sun's altitude. This is based on the archaic ring dial of Aryabhata and Varahamihira in which the sunlight enters a hole in the breadth of the ring and falls on the inner concave surface on the opposite side, indicating the sun's altitude.

Tamil Sources

Navoi Sattiram which is a part of the McKenzie collections in the Madras Archives appears to be older than *Kappal Sattiram* of Tarangambadi at least by a century and a half. This is a Tamil palm leaf manuscript. It has been edited by Soundarapandian in 1995.[6] This is primarily astrological in nature and refers to boat building. It was written by Trikuta Nambi. It has a couple of stanzas each relating to *Maram Vetti Vasthu Shastra* (the science of *vastu* or architecture relating to timber felling), *mara kala muhurta* (auspicious time for initiating building of a ship) and *Era edukka* (laying the keel) which goes into details of timber quality such as physical defects, the colour of a freshly cut section of the log and stains.[7]

The *Kappal Sattiram*, a Tamil treatise, is of 1620. It was written in Tarangambadi and is more useful than *Yuktikalpataru.*This has been critically edited with introduction by T.P. Palaniyappa Pillai.[8]

Kulatturayyan Kappal Pattu[9] is a work available in manuscript form written on palm leaf. The work was designed to eulogize Kulathur Manikanda Ayyar, a Brahmin merchant operating in the region of Quilon-Trivandrum coast late in the eighteenth century. Kulathur was located near Kollimalai, otherwise called Accan Koyilayyan near Quilon in south Travancore. This ballad gives us some idea of the wood used in shipbuilding and also about the sails. The craftsmen went in search of suitable trees in the Western Ghats, near Ponmalai, across *Podigai* hills near *Sandanakkadu*. The wood chosen by the craftsmen consisted of *Vembu (Margosa), Ilupai, Punnai, Krimaruthu, Sirunangu, Aini, Kongu* and *Naval*. These were for the floating bottom. They chose *Venteak* and teak for side planks and supporting beams, and used ropes and coir for binding the planks. The planks were strengthened by metal plates made of gold and nailed with rubies (*pasumponnai ellamorupal tagadakki reddina anitariyittu*).

Glues were applied to the sails lest they should change direction on their own. Since the ship was propelled by wind, sails were used. The *Kulatturayan Kappal Pattu* speaks of sails used for the propulsion of the ship. The ship had several sails (*Tiralmiru Irettu Pamaram nirutti*).

Cala Vattu Pattu of Nagapattinam-Nagore distinguishes timber as masculine, feminine and eunuch based on log girth.[10]

Islamic influence

The making of instruments did not interest Indian astronomers, but interaction with Islamic astronomy paved the way for a substantial change of attitude. An outcome of this interaction is the introduction of the astrolabe into India by Al-Biruni in the eleventh century.[11] Muslim scholars from Central Asia migrated to India after the establishment of the Delhi Sultanate. These scholars brought astrolabes with them and used them in India for calendrical and astrological purposes. By the middle of the fourteenth century the astrolabe was well known among the Muslims of north India. There is a long account of the astrolabes manufactured under instructions from Firuz Shah in the *Sirat-i-Firuz Shahi*, an anonymous chronicle of the period.

Medieval Arab Accounts of Navigation and Shipbuilding

Works of Ibn Majid and Sulaiman al Mahri constitute the earliest Arab accounts of sailors themselves. The sources for the work of Ibn Majid go back to the twelfth century. He was the first to write a work on navigation drawing on his own experience besides the wisdom collected from earliest generations. He belonged to a family of navigators of the northern Indian Ocean. He lived in the latter half of the fifteenth century. He had travelled extensively between East Africa and Malaya including the Red Sea. His best text written in prose is *Fawa'ld* (AD 1475). He was known to Portuguese chroniclers like Barros and Castanheda as Malami Canaqua. Gabriel Ferrand holds that there is no doubt that Ibn Majid was the Malami Canaqua, the master of astronomical navigation.[12]

Sulaiman Al Mahri lived sixty years after Ibn Majid. He wrote two of his important works, *Umda* and *Minhaj* around 1511, in prose. Though his works are not as profound as those of Ibn Majid, they are better organized and orderly compared to the works of Ibn Majid. They reveal the maritime wisdom, skills and techniques of the fifteenth and sixteenth centuries. He deals with the nautical and astronomical concepts, principles and rules that were to be followed.

Work in Turkish Language: *Muhit* of Sidi Celebi

Sidi Celebi was a Turkish writer and admiral who had been in-charge of Sulaiman the Magnificent's Indian Ocean fleet (Turkish Ottoman fleet). He wrote the *Muhit* in 1554 at Ahmedabad in Gujarat in Turkish during his enforced stay there after the dispersal of his fleet in 1554. This work was chiefly a translation from Arabic of several of the works dealing with navigation.[13] He compiled with care the oral and written records of the Arabs, Persians and the Portuguese navigators. His work translated in English appeared in the *Journal of the Asiatic Society of Bengal* between 1834 and 1839. He deals with the sky and the stars, solar and lunar years, divisions of the compass, rhumbs and tirfa, sea-routes, wind on the high seas, nautical measures, star altitudes, inter-port distances, winds and monsoons, maritime route details, and sea dangers including cyclones, in ten chapters.

Medieval Accounts of Indian Navigation

Sailors (*tandels* and *malmis)* in the region of Lakshadweep-Malabar use some manuscripts of sailor guides which go back to the past three or four centuries. They are in the form of *Rehmanis* and *Roznamas*. *Rehmani* is a Persian word meaning a sailor's manual or guide book of instructions meant for sailors of later date. *Roznama* is a daily diary or log book of an actual voyage, on a day today basis, giving details of the day's position of the vessel, weather conditions and other events of importance on board.[14] The Arabic works of Ibn Majid, Suleiman al Mahri and Sidi Celebi are *Rehmanis*.

Navigational material gathered from Lakshadweep, Malabar and southern Tamil Nadu have been written in *Arvi* and *Arabu*-Malayalam (Arabic using Tamil or Malayalam languages). One such *Arabu*-Malayalam text is in the possession of a Thangal Family in Kavratti.[15]

Kannakiyum Cheermakkavum[16] is a ritualistic song related to the worship of Kannaki in certain Sreekurumba shrines of Malabar. It provides some narrative accounts of shipbuilding. The song speaks of the selection of tree made by a *Tachan* who went to the mountain to build a vessel for a goddess. He made a keel (*pandi*) from the tree he felled. He arranged nails and copper plates and conducted the auspicious *ganapathipooja*. The length of the vessel was 40 *kol*. It was divided into forty compartments. Then several ribs (*kal*) were fixed at a reasonable distance. The back was covered by planks, and further joined by nails. The *Tachan* fixed the *aniyam* (stern). The vessel had seven decks (*Tattu*). Then he erected the *kombu* and *tandus* (oars) for the *tandalas* (boatmen) Further the *Tachan appended achukan* (rudder) and anchor (*namkuram*).

Other equipment like *kambakkayar* (rope), pulley, *pamaram* (mast) and sail cloth (*paya*) were gathered. He applied *pantham* for caulking to prevent the entry of water inside the vessel or *kalli* (compartments). *Kuttipuja* was performed by the *Tachan* by sacrificing a number of fowls. The *Tachan* was given silk clothes, decorated bangles and a jewelled ring by the goddess. *Karivaka, Venteak, Pilavu, Karimaruthu, Punna* and *Aini* were used profusely for shipbuilding in Malabar. The English East India Company monopolized the teak forests in Malabar by 1810.[17] We can assume that the Tamil sources and *Rehmani* would reflect the indigenous shipping and shipbuilding before the arrival of the Europeans though they were composed later than the sixteenth century.

Indigenous Shipbuilding at Calicut (Beypore)

Beypore, south of the present city of Calicut, was the most famous centre of shipbuilding before the arrival of the Portuguese. It is situated on the bank of river Chaliyar known also as Beypore River. Through the two branches of river Chaliyar which join together in the Nilampur Forest Beypore is connected with the Nilgiri mountains and Wynadu ghats. River Kallai joins Chaliyar River before it opens itself into Arabian Sea and so Beypore gets access to river Kallai too. Further through river Kadalundi Beypore is connected with Kadalundi known for the best timber. Thus Beypore has access to a large timber producing area.

The well known types of vessels known as *uru* and *odam* have been built in Beypore from time immemorial. They are famous wooden vessels. Even the Arab literature has reference to the excellent timber available on the Malabar coast and to the skilful carpenters who built exquisite vessels. *Urus* made in Beypore are unique. They are built entirely of timber fastened by coir rope and caulked with a special glue made of animal fat, calcium, and *punna* oil and wooden nails.

Timber was used for the construction of vessels at Beypore .*Karimaruthu* (*Terminalia crenulata*), *karivaka,* Benteak (larger *Stroemia lanceolota*), *Pilavu* (*Artocarpus integrifolius*), *Pali* (*palquim elepticum*), *Punna* (*Caleophyllum indophyllum*) teak (*Tectona grandis, Ayini* (*Artocarpus birustus*) and *Cini* (*Samnea saman*) were some of the species of trees used for shipbuilding. Timber for the shipbuilding at Beypore was obtained chiefly from Nilambur forests. The ships built on the Coromandel coast during the tenth and eleventh centuries were all timber-built and had two or three masts, according to the size of the vessel.[18]

The Felling and Transportation of Suitable Trees

After deciding the size of the vessels to be built, carpenters, under a *muppan* (elder), went into the deep forests to identify the appropriate trees. The wages for them were disbursed to their families directly every week. During the lean season when there was no work, they used to take advances from timber merchants. The felling of trees was done only during the appropriate season of the year, depending on the position of the moon. On an auspicious day and time (*muhurta*) the desired tree was felled with a ceremonial saw.

Trees for shipbuilding should be cut only when they are mature and in the proper season. If the trees are not mature, the timber would rot or create changes in the work by twisting or shrinking and opening up joints. If the trees are not mature, there will be too much sap and so the timber would be corrupted and rotten. All trees do not mature at the same time. Some kinds of trees mature differently in different places. They will mature faster in warm areas than in cold regions.[19]

The season after which trees are mature, and ought to be gathered, is after they have given their fruit, and, all their virtue and strength gathered in, and they are resting before giving fruit anew. The cause of this change is the presence and absence of the sun which approaching the trees with its own movement warms them, and with its heat gathering the nourishment of the earth to them, they conceive their fruit in proper season and they grow. The trees growing in tropical regions are harder and more durable than those grown in other areas on account of the proximity to the sun and the warmth of the sunrays. While fixing the season for felling the trees it is necessary to have regard for the movement of the moon since it is closer to earth than other celestial bodies. It is considered proper to fell the trees in the waning phase of the moon. Greater care has to be taken in cutting the trees for shipbuilding than for building houses because corruption of the timber for a house is of little consideration in comparison to the rotting of a plank of a ship. Intrinsic dampness of the timber is to be dried before the timber is used for building ships. If the tree is cut to the middle of its heart, all the superfluous dampness may drain away and being dried, it may become free from rapid corruption. The felled trees are left for many days either in the field or in the shipyard or in salt water, and they may not be worked until after it is known that everything that may be feared has happened to them.[20]

A track was prepared for transportation. Big logs were dragged through by male elephants till the *mara elu* (wood track). Naikan and Paniya tribesmen constructed the *mara elu* to drag the felled trees selecting the route and deciding the slopes. *Mara elu* was connected with a bigger *elu* used by many such contractors. Sometimes the *ana elu* (tracks used by elephants) was modified for the purpose. This was very economical. The felled trees were dragged to the river by male elephants and were floated on the river. A number of them were tied together with hundreds of dry bamboos for keeping the trees floating. This mode of transportation was called *therappam*. Sometimes these trees were made to float with the help of *Tonis,* canoes or punts propelled by paddles or poles respectively. *Therappams* were prepared, smaller in size with fewer logs in the upper river with shallow water and fast currents. Two or more such *therappams* were joined together in the lower reaches. The *therappakar* moved on the river day and night with stops only for food and other requirements. Large number of *makkanis* (hotels) sprang up in summer along the sandy banks especially near important ferries to provide food to the *therappakar*. *Tonis* of *therappam* were used to carry goods on their way back from the shipbuilding or curing centres.[21]

Stages of Shipbuilding

Traditional carpenters worked in the shipyard. They had their experienced headman, called *mestris*.[22] The *mestri* decided the size of the frame and planks as dictated by his own memory. In fact there was no prepared chart or drawing. The carpenters worked under the *mestri* with great discipline.

Keel (*eravu* or *pandi* or *patan*) and keelson, the stem and stem posts, the lower ribs, the side-planks permanently below the water level, the cross beams, the masts and spars are the important parts for which the greatest care was given to maintain the quality of timber. Selection of planks for the upper level side above water-level, the inner decks, cabins and platforms permitted some flexibility.

An experienced *mestri* or master carpenter, after deciding the size of the vessel to be built, cut the planks as required depending on the plan he had in his memory.

The Irippu (seat)

Wooden seat upon which the keel of the ship was to be made, was made strong enough to support the entire weight of the vessel. It was made of two wooden *makkanis* (big wooden pieces) on two sides and then keel was laid.

Keel

The keel is the foundation or backbone of a ship and made first. It should be a straight and single piece of timber without any curve. Keel laying was an important function and was done on an auspicious day chosen on the advice of 'kanakkan' or astrologer. Coconuts were broken to ward of evil spirits. Betel leaves and nuts were distributed to guests invited to witness the ceremony.

Keel is the most important part of the vessel which gives strength and stability to it acting as the backbone, besides being the single massive timber. The length of the keel is the same as that of the vessel. A keel of 110 feet has a 16-inch width and thickness. The preference is for *Karimaruthu* wood (*Terminalia crenulata*). The procurement of a single piece of *Karimaruthu* of 90 ft. was a difficult task. Usually the length of the pandi or keel was limited to 90 ft. It is rather difficult to get one piece of timber of this size without physical defects. Two beams known as *Aniyam* (stern) and *Amaram* (stem) posts were joined to it at both ends. The stern post was connected in such a way that the angle against the water surface was greater than the angle between the stern post and the water surface when the ship was launched. Two or three timber pieces of the same wood are used by joining the planks together. The keel planks are joined by interfingering tongue and groove scarf joining and fastened with wooden pegs. Joints in the keel are fastened by pantham – a resin of *Canarium sprictum*. The keel was covered by a beam, its shape depending on the shape of the bottom of the ship.[23] The carpenters fixed a plank called *ottupalaka* on either side of the keel. The position of *ottupalaka* towards *aniyam* and *amaram* is 0° and in the middle 45° C. Before fixing the *ottupalaka*, on the keel, the carpenters kept cotton immersed in a gum mixture of *pantham, Punna* oil and neem oil for waterproofing. *Ben teak* (*Lagerstroemia lanceolota*) which swells in water, thereby making the keel water tight, was used.[24] *Kappal Sastram* recommends *Vembu, Ilupai, Punnai* and *Naval* as the ideal timber for *keel,* while *Karumaruthu, Sirutekku, Sirunangu, Ayini, Karu-nelli, Kongu* and *Vengai* are suggested in the *Kulatturayyan Kappal Pattu.*[25] The general trend was to use *Karimaruthu* and in its absence *Punnai, Ilupai*, and seldom *Seerani* (*Puvarasu*) or *Vembu.*

The length of the keel depended on the size of the ship. A ship having a tonnage of 500, should have a keel of 90-5 feet length. The calculations are as below:

- Tonnage 500.
- Keel length. 90-5 feet
- Breadth 35 feet
- Height 19 feet

Hull

Hull of a ship consisted of keel and ribs covered by planks. The hull could be of different shape with a flat or V or round bottom. Once the keel was fixed the ribs (*manikkal*) were prepared. It was a leg-like support fixed on the keel. A vessel of 110 feet required 50 ribs on each side. The length in the middle was 31 feet reducing towards either side while the length in the sides was only 2 feet. The width of the ribs was 10 and 8 inches in thickness. The distance between two ribs was 3 inches. A *mattam* or model was prepared before shaping the ribs. Ribs were usually made by joining two pieces. *Manikkal* (ribs) or *Mallakkals* (side legs) were placed at reasonable distance taking into account the size of the vessel. These *mattakals* were fitted on two sides in a pair. Different types of legs were used for various purposes such as (a) *otharkal*, leg to tie the stay bar, (b) *buoykkal*, leg to tie the anchor in the stern, (c) *pathikal*- to spread the *panthal* or cover as roof, (d) *peelikkal*, used for additional sails. The *manikkal* and planks were fastened together with coir.

In the past, sewn boats were made in India, usually called *masulas.*[26] After fixing the ribs on either side ten planks of 4 inches width were placed to strengthen the ribs.

Outer Planking or *Kakaorayam Cherkkal*

The carpenters made 'V' shape on one plank and to that they inserted the other piece. The 'V' shape was called *kakaorayam* since it looked like the open beak of a crow. Bending of the timber according to the requirements for building of ships was done by the use of *mara enna* (oil extracted from trees). *Vep* (neem) oil or *cheeni* oil was applied on the planks. Then they were heated up mildly. There was another method for the same. The planks were besmeared with a layer of mud found on the river banks. Mud of this type was greasy and paste-like with some special properties. The moisture from the mud was allowed to be absorbed. The planks were kept as flat on two wooden logs at the ends. Then the plank was heated till the slush gave out steam. The steam entered in the body of the planks and made them elastic. The planks became pliable. One end of the plank was then inserted into a long vertical slit cut in a thick wooden stump. The other end was fastened by strong rope lashed to a big tree. The rope and the plank were pulled according to the desired curve. The slanting position of the middle pole helped in bending the plank to the required shape of the hull.

Stern (Aniyam) *and Stem* (Amaram)

The stem was the extreme front of a ship. Its size depended on the shape of the ship and usually had a 7 inch slant. A vessel of 110 feet had an *amaram* of 35 feet height and an *aniyam* of 24 feet. These were fixed before the ribs were finally placed. *Amaram* was at the extreme back of a boat. It was controlled by a sculler or *Amarakkaran.*

The *cukkan* (rudder) was fitted into the *amaram* with the help of a bronze ring in such a way that it could move in any direction. *Cukkan* was used to control the direction of the vessel by steering it. It had the height of the stern. Its width increased towards its base.

Sails (*Paya*)

Sails were made of thick cotton fabric. Three sails of 150 m, 200 m and 40 m are used in some types of vessels. The big sail was fixed in the front. The sail cloth called locally in Malabar as *Payathuni* was stitched and prepared in different sizes according to the shapes required: triangular, mizzen, lateen, square, etc. Ludovico di Varthema says: 'the sails of these ships are made of cotton, and at the foot of the said sails they carry another sail, and they spread this when they are sailing in order to catch more wind, so that they carry two sails where we carry one'.[27]

Mast (*Kombu* or *Paymaram*)

The height of the mast depended on the size of the ship. It was erected on the keel as its base, where a hole called *pandi kuzhi* was designed for it. The sail cloth was fastened on it. The mast was fixed on a pace which was ⅓ of the keel length. Ships would have one, two, or three masts. If the ship had only one mast, it was fixed at the centre.

Two-masts, *kombu,* of 60 feet and 45 feet in length were used in vessels known as *Uru.* A mast of 60 feet required a 1.8 foot circumference while that of 45 feet needed 1.4 feet width. *Punna* was usually chosen for the mast. A yard known as *pariman* on the Malabar coast was used for tying the sail. It used to be 128 feet long. It was made of *punna* tree. A pulley (*kappi* in the local language) was used to spread the sail. *The kappi* was usually made of the wood of the jackfruit tree.

Representations of ships from the Ajanta frescoes were interpreted by Griffiths. One of the ships was a sea-going vessel with a high stem and stern having three oblong sails attached to three masts.[28] Some Andhra coins of the second and third centuries AD depict large sized ships with two masts each.[29] Similarly, some coins of the Kurumbar or Pallavas of the Coromandel coast have representations of two-masted ships propelled by oars from the stern. The Kurumbars were a pastoral tribe living in associated communities some hundred years before the seventh century, the area from the base of the tableland to the Palar and Pennar River some of whom were engaged in sea-borne trade.[30]

Nails for Shipbuilding

It was generally held that Indian shipwrights did not use nails to join the planks of a ship in the period before the arrival of the Portuguese. Wooden pegs and coir were used lavishly in joining and tightening the planks. Varthema, however, mentions that an immense quantity of iron nails was used in shipbuilding in Calicut in the early decade of the sixteenth century: 'And when they build the said vessels they do not put any oakum between one plank and another in any way whatever, but they join the planks so well that they keep out the water most excellently. And then they lay on pitch outside, and put an immense quantity of iron nails'.[31] In fact there were old directives not to use iron nails on the ship since rock formations in the sea have magnetic areas which may cause difficulties for navigation.

Anchor (*Nankooram*)

Granite stone carved specially in a square shape with a sharp edged wooden piece in the middle tied with coir (*kal*) was used in Malabar as anchor. It was fixed in the sea bed when thrown from the ship. Marble pieces were used as anchors for the vessels built in Calicut. Anchors of this type, 8 palms long and 2 palms broad and thick, were tied to the vessel through two large ropes.[32] Anchor rope was made of coir 6 to 7 inches thick and was called *alath* or *vadam*. This was needed for operating the anchor.

Coating and Outer Treatments

Water proofing was done after completing the planking. Cotton treated by specially made mixture of *punna* oil and *pandam* was inserted in the gaps between planks. Fish oil or any vegetable oil was applied up to the water level for seasoning the vessel. Another mixture, called *cherivi* with lime and *punna* oil was also applied. This mixture protected the vessel from shipworms.

The Portuguese writers of the sixteenth century mention materials such as *Galagata* and *Saragusta,* made indigenously, for the treatment of the vessel.

Galagata (*gualagualla*) or bitumen was made of three materials, namely virgin lime, fish oil, and linen. If fish oil was not available, gingely oil or any other oil was used.[33] This sealant served mainly as a protection against shipworms, which could not penetrate from outside, their teeth being dulled by lime.[34]

Saragusta

It was made of four materials, lime, *allcatrão de breu*, fish oil, and linen.[35] The material prepared out of this was used in India to treat the seams of ships.[36]

Caulking (*Panthavum Paruthiyum*)

Kalpath was the term used for the caulking of a ship. It was also known as *panthavum paruthiyum* in view of the articles used for caulking. *Pantham*, a resin taken from certain species of tree, was essential to embalm the ships to protect them from worms and to prevent leakage. *Paruthi* meant cotton. Usually *punnakka enna* (oil of *punna* or pine tree) along with cotton and coconut fibre (*chakiri*) was used. Cotton and coconut fibre were soaked in the oil of *punna* and applied to the chinks and strongly hammered. Some scholars are of the opinion that the *kalpath* was not needed for ships made on the Malabar coast, because Indian shipwrights carved each piece of timber according to the shape of the hull suggested to them. Each plank was filled to its neighbouring piece until a perfect joint was effected. Here, waterproofing as required for vessels made in Europe was superfluous.[37]

Chopra (Embalming a ship)

After caulking had been done inside and outside, *chopra* or embalming was done. It was applied from the bottom to the water line with a white composition. This mixture was made from the *pantham* (resin), *dammar,* fish oil, and burned lime (*chunna*). All these items were boiled and carried upward in a wide sweep on either bow. Above this, a coat of paint was applied which separated the pitched bottom from the sides. *Chopra* protected the wood from termite, prevented leakage, tightened the caulking materials into the chinks, prevented the ship from decaying, and gave a fresh appearance to the ship. *Chopra* coating was done once in three months.

Shipbuilders on the Malabar Coast

The shipwrights on the Malabar coast were mostly Hindus known under the generic name *thachan* or *asari*. Their knowledge is hereditary. The shipwrights of Beypore are called Odayees, forming a subgroup of the carpenters. They belonged to the Viswakarma group of Malabar. They are inhabitants of Calicut, Kallayi, Beypore, Panthalayanikollam, and Madayi. They made a livelihood from the construction of sail boats, *urus,* booms, *kotia* and *sambuk*. Odayees of Calicut traced their origin to the times of the legendary Cheraman Perumal. They used to call themselves Cherman Odayees. As per the Settlement Register, Odayees of Calicut and Pantalayani have been living as tenants of the *jenman* land belonging to the Zamorin's family.[38] They used no pre-drawn design of any ship except what they had in their memory.

Calicut (Beypore) was considered to be the best centre for shipbuilding during the period before the arrival of the Portuguese. Calicut was reported to have good timber in great abundance surpassing the supply of timber in Italy. The availability of good variety of timber suitable for the building of ships in the interior places like Nilampur and the possibility of transporting it from the interior to Calicut through riverine traffic are well known. Kallayi is famous for the treatment of timber in water. Kallayi River was the major means of bringing timber from the interior to the areas of shipbuilding.

Tonnage

Pliny gives information about the tonnage of some ships in the Indian Ocean. According to him, Indian vessels had a tonnage of 3,000 *amphorae*, the *amphora* weighing about a fortieth of a ton.[39] Tonnage varied from three hundred to four hundred butts as observed by Ludovico di Varthema.[40] In the days of prosperity, i.e. before the arrival of the Portuguese, the shipyard at Calicut built keeled ships of one thousand to one thousand two hundred *bhares'* tonnage.[41] It was observed by Ludovico di Varthema that the shipbuilders at Calicut did not put any oakum between one plank and the other but, they were experts in planking the ships perfectly watertight. They used a lot of iron nails according to Varthema though the general observation of visitors is that the use of iron nails for the building of ships was not common on the Malabar coast before the arrival of the Portuguese. Thus, for example, Duarte Barbosa who was in India since 1503 while speaking about shipbuilding in Calicut before the arrival of the Portuguese firmly asserts that the ships were built without iron nails. He adds: '. . . the whole of the sheathing was sewn with thread, and all upper works differed much from the fashion of ours, they had no decks'.[42] The sails of the vessels made in Calicut were of cotton and there was always an extra sail besides the main. This was spread when the navigators wanted to harness more wind. This was something different from the practice in Europe where only one sail was used.[43]

Indigenous cargo ships plying the Malabar coast were sometimes of six hundred tons. A contemporary Portuguese chronicler provides mention of a ship belonging to Mammale Marakkar and Cherina Marakkar of Cochin which carried 7 elephants from Ceylon and 300 armed men on board.This ship was on its way from Cochin via Calicut to Gujarat around 1500. It was of 600 tons.[44] We have reference to another huge ship, *Meri* by name, plying between the Malabar coast and Mocha. It carried many families of the Muslims from Calicut to Mocha. It had 260 soldiers on board in addition to mariners and a rich cargo. It was sighted near Mount Eli.[45] These two incidents give an idea of the tonnage of the ships plying by the Malabar coast in the first decade of the sixteenth century.

Types of Ships

Several types of vessels were found in Calicut during the period before the arrival of the Portuguese. The flat-bottomed vessels known as *Sambuk* were made in Calicut. Vessels with bottom shaped like those of Italy were also employed in Calicut for navigation. They were called *Kappal.* Another variety was called *Parao* (*Prahu*) measuring ten paces, made of a single piece of wood. Boats of this type were propelled by oars. The mast for such boats was made of cane. Similarly ferry-boats made of one piece of timber called *Almadia* were also built in Calicut. Another sort of vessel made of a single log was propelled by oars and sails. This measured twelve to thirteen paces in length. All vessels made of single piece of timber had a very sharp opening and were fast in movement. They were called *Chaturi* and excelled Italian galleys, *fusta* (foist) or brigantines. The corsairs must have used vessels of this nature.[46]

Ludovico di Varthema of Bologna who was in Calicut in the early part of the first decade of the sixteenth century writes:

> As to the names of their ships, some are called *sambuchi* [sambuk] and these are flat-bottomed. Some others which are made like ours that is in the bottom they call *capel* (*kapal*). Some other small ships are called *parao* (*prahu*, prow), and they are boats of ten paces each, and are all of one piece, and go with oars made of cane, and the mast also is made of cane. There is another kind of small bark called *almadia* (*al-ma'adiya*, ferry-boat), which is all of one piece. There is also another kind of vessel which goes with a sail and oars. These are all made of one piece, of the length of twelve or thirteen paces each. The opening is so narrow that one man cannot sit by the side of the other, but one is obliged to go before the other. They are sharp at both ends. These ships are called *chaturi* [*shakhtur*], and go either with a sail or oars more swiftly than any galley, *fusta* [foist], or brigantine.

There are corsairs of the sea, and these *chaturi are* made at an island which is near, called Porcai [Porrakad].[47]

Based on the mode of construction, vessels were classified as (a) *kattamaram*, (b) dug-outs, and (c) plank-built. *Kattamaram* was built of three or four logs tied together. Dug-outs were cut out of a single trunk usually of the mango tree. This was similar or identical to *monoxylon* dug out from the trunk of a tree. Dug-outs ranged from one man *toni* to eight men crew of *odams*. The *colonidphonta* was a large ocean-going ship.

BATIL, OR BATEL

It was a two-masted sailing vessel. A decorated wooden stern was its most attractive and striking feature. It had only one deck and the vessel was made of teak wood.

BOOM

It had three masts with three sails. It had no cabin though there were three decks with splendid platforms at prow and stern. It was usually built at Beypore for Arab merchants.

SAMBOOK

Sambook was also built at Beypore for the Arabs with a more flat stern. Forty to forty-five persons could be accommodated in this vessel. Its planks were fastened together.

KOTIA

The *Kotia* was built at Beypore for the merchants from Kutch and Gujarat in general. The peculiar architectural feature in the stern was its speciality.

Berik and *dungi* were also small vessels built at Beypore for merchants from Gujarat. *Berik* built in Malabar was not round shaped in the back, but square and slant in the front.

PATAVU (*PARAO*)

It was also built in Malabar for the local transportation. Huge cargos were carried by the *patavu*. A typical *patavu* is of a keel length of 100 feet, a beam 21 feet across, a depth of 4 feet and a tonnage of 150.

In Malabar it was built of teak planking, *margosa* (*Azadirachta indica*) frames and masts and spars of *punnai* (*Caleophyllum indophyllum*). Artocarpus nobilis timber was also used in place of teak.[48] *Pattemari* was used for transporting heavy cargo.

Launching a Ship

Launching the ship was a happy ceremony for the owner, merchants, passengers, shipbuilders, labourers and crew. The owner for whom the ship had been built invited his friends and relatives to grace the auspicious moment of the first launching of the ship. All the guests brought a coconut each and stood in a row in front of the vessel. They went around the vessel by knocking the coconut on the hulls and side planks. Then they broke the coconut on a stone to ward off evil spirit. *Ganapathipooja* was performed under the leadership of the *mestri*. The guests used to give some presents to the *mestri*. The owner of the ship gave a *Mothirakkani Veshti* to the *mestri* and cloths and money to other workers.

The *dhawaring*[49] technique was used to launch the ship. According to this method, green coconut leaves were spread on the sand where the vessel stood. Pieces of round coconut timber were placed on the leaves as rollers. The rollers did not sink on the sand due to coconut leaves. Supports were placed on the sides to prevent the ship turning sideways. The huge ship weighing hundreds of tonnes was pushed by using a pulley mechanism called *dhawar*. The ship was made to move slowly on the rollers to the sea.

Dhawar consisted of a removable stand with two pieces of huge wooden rollers. There were two movable thick and long posts called *kai* in the middle of the rollers. A big beam was placed horizontally in between the wooden rollers. This was called *pakku*. A *kamba* (big rope) was connected with *kappi* (pulley). One end of the rope was tied to the stern of the ship. When the workers turned the wooden *kais* in a circle, the rope tightened with the pulley and the ship moved on the rollers to the sea without applying energy manually. A small *dhawar* was used in the stem of the ship to lift the anchor.

Dhawar was operated for launching the ship as well as drawing the ship on to the coast by Mappilla *Khalasees*. *Khalasees* constituted a traditional group of Muslims engaged in risky operations under the supervision of a *moopan*.

Navigation

Navigation from Calicut was governed by the monsoon. It was possible only for eight months in the year, namely from September to April. Navigation is not possible from May to August on account of the fury of the South-West monsoon. The vessels crossed the Cape of Comorin and entered another course of navigation during this period. Very heavy showers were common during the months of May, June, July and August in the period before the arrival of the Portuguese.[50]

In view of the remarks from R.H. Ellis, a British officer in 1923 that there should be course on modern navigation, a textbook called *Navika Sastram* in Malayalam was published in 1939 for teaching navigational history to students at Amini, in the Laccadives. Mariners on the Malabar coast made use of the accumulated and inherited wisdom about celestial bodies, winds, animals and flora and fauna for navigation. The wind is the force that gave ships propulsion in the deep seas when the sails were utilized.

The Indian mariner threw into the sea during the prevalence of calm,[51] a ball of ashes kneaded together by water. As it slowly sank, it separated, leaving a long broad tail, like that of a comet, behind it which wafted away in the run of the current, making a line of direction apparent to an observer standing a little over the surface. It is reported that a similar device was used by the Portuguese in the sixteenth century.[52]

The most important task of a navigator was to find the direction in which his vessel was moving towards the horizon, by observing the sea currents, winds, sun, moon and stars, birds, and sea weeds. This observation of the sun, moon and stars was conducted with their naked eyes. The navigators by observing the wind could point out its origin and the direction in which the vessel was moving.

The sailors used animals and birds to find the direction of the vessel. Two seals (*c*.2500-1750 BC) discovered from the DK area of Mohenjo Daro and one graffiti have representations of sea-going ships. There are two birds, forward and aft which seem to be *disha-kaka* used for finding direction 'It was common to carry such birds aboard, because their infallible flight towards the land when released helped the mariners in locating the direction of the land'.[53] The Biblical Noah kept a raven and a dove in his ark and let then out at intervals to find out dry land.[54] Crows were carried in the ships in the past to find the direction. They were called *disha-kaka*. Pliny in his description of Taprobane (Ceylon) writes:

> The Sea between the island of Ceylon and India is full of shallows not more than six paces in depth, but in some channels so deep that no anchors can find the bottom. For this reason ships are built with prows at each end, for turning about in channels of extreme narrowness. In making sea voyages the Taprobane mariners make no observations of the stars, and indeed the Great Bear is not visible to them, but they take birds out to sea with them which they let loose from time to time and follow the direction of their flight as they make for land.[55]

A monkey (known in Malabar as *Kuttithevang*) found in the forests always sits facing the direction opposite the sun. Monkeys of this type were carried on ships as pet animals. Similarly, birds kept on board were released from the ship to find the direction of the land towards which they flew. Navigators were able to identify their location at sea by observing the fishes, the colour of water, sea-crows and seagulls, flora and fauna, debris, fixed stars and the horizon.

Finding the Direction through *Viralkanakku*

Kau Nila (status of the *kavu*, i.e. Pole Star) or *Kavu Kanakku* is calculated through the use of *virals* (fingers).This star measure is known as *viralkanakku* among the navigating seamen of the Coromandel and the Gulf of Mannar coast. Locations of coastal port sites and islands are estimated from the zero altitude position of the Pole Star at intervals of ⅛, ¼, or ½ *viral*.

Kanyakumari is placed at a location of 1.5 *virals* altitude. Vizhinjam is at 1.75 *virals*. It was the battle site among the Cholas, Cheras and Pandyas during the tenth and eleventh centuries. During the reign of Rajaraja Chola it was believed to be naval battle site on more than one occasion. Southern Kollam is placed at 2 *virals*. Kayamkulam is located at 2.25 *virals*. Katturkadavu north of Kayamkulam is placed at 2⅜ *virals*. Purakkad is situated in this barrier island. Ambalapuzha is placed at 2.5 *virals* while Kochi Orumukam is at 2⅝ *virals*. Kodungallur is situated at 3 *virals* altitude, Ponnani port at 3.25 *virals,* Tanur at 3⅜ virals, Chaliyam near Beypore is at 3.5 *virals,* Korapuzha at 3.75 *virals,* Dharmapatnaturuttu at 3⅞ *virals,* Kannur at 4 *virals,* Elimala is at 4⅛ *virals;* Nileswaram on the Kavvayi backwaters at 4.25 *virals*; Kasargode at 4⅜ *virals*, Manjeswaram and Kumbala at 4.5 *virals*; Mangalore at 4⅝ *virals*.[56]

Measurement of the Depth of the Sea

The depth of the sea was estimated according to the colour of the water. Green or a pure blue colour indicated deep waters. Progressive reddish brown colour shows shallower waters. The simple unit of depth used among the Indian seamen is the height of an average adult man or the span of length between the extreme ends of horizontally stretched arms at shoulder height. The term used to indicate the span of length of the arms is *maar* or *pagam* in Malayalam and Tamil. It is equal to 4 cubits (*Muzham*).

Winds

Winds did play an important role in navigation since it provided natural energy needed for propulsion of the vessels. These winds themselves helped the mariners to find their direction. The seamen had the knowledge of different categories of wind. An Arab pilot Abu Haneefa Dianuri in his work on nautical sciences mentioned twelve types of winds and another author Avasya Ka Cursin identified sixteen categories of winds. Seamen of Malabar coast knew eight types of winds.

1. *Thekenkattu* (southerly wind) . It starts from the south and then passes on to the north-west. It occurs in the month of *Edavam* (June). It is not favourable for navigation.
2. *Vadaken kattu* (*katchan kattu*, northerly wind). It originates from in the north and blows to the south. It is also called *Makara kachan*.
3. *Karakattu* (wind from land or easterly wind). It begins on 20 *Thulam* (November) and lasts upto the beginning of *Dhanu* (January). It blows from the east to the west. It makes it difficult for a vessel to steer towards the land. It begins after midnight

and continues upto 1 in the afternoon and never blows 10 miles into the sea from the coast. The sailors go to sea at 3 or 4 a.m. utilizing this wind. It is easy to navigate since the wind comes from the land. This helps smooth sailing during the period from December to January.

4. *Purathkattu* or *Padinjaran vadai* is the westerly wind. It begins after 1 O'clock in the afternoon and goes on upto midnight. The westerly wind blows from the sea to the land and sailors will avoid the time of this wind from the coast to sea.
5. *Kalavarsha kattu* (south-west monsoon wind). It blows from the southwest to the northeast during the months of June and July. Navigation on the western coast is difficult during this period.
6. *Thenkara kattu* (south-eastern wind). It blows to the west during the month of *Medam* (May). If the sailors are at sea in their ships it will be difficult to control the vessel and they could drop the anchor and wait until the wind stops.
7. *Vadorth kattu:* It blows during the months of *Dhanu* to *Meenam* (end of January to the beginning of April). Boats can easily come to the mud banks then.
8. *Vattakara Kattu* (north-easterly wind). It blows in *Dhanu-Makaram* (November to January) period . The vessels cannot take position from the north-east to the south-west to reach the shore when this wind is on.

Sea Currents

Currents have a great impact on navigation. Therefore the navigators should be familiar with the direction of currents. They are known in Malabar as *neer baloo* and *ilakkam.* The combination of winds and currents can be fatal if one did not know the dynamics of them. The following are the different currents:

(a) *Thekken neer:* When it joins with *Kalavarsha kattu* it creates sea squalls. The waves swell.

(b) *Vadakkan neer.* This is from the north to east. It takes place between *Kanni* and *Meenam* (October to April). When *Vadakkan neer* joins *Thekken neer* high waves are produced.

(c) *Kara neer* or coastal current: *Vadakkan neer* leads to *Kara neer*. It goes from east to west near the sea-shore.
(d) *Pura neer* (westerly current): The strong undercurrent moves from west to east.
(e) *Thekkupadinjaran neer* (south-westerly current): This is the famous *Dhamma neer*, because its *Dhama* slush will come up from the bottom. The colour turns red.
(f) *Vadorth neer*. It has two different currents. The *melvellam* (top current) moves in the direction from west to the east and *adivéllam* (undercurrent) from east to west.
(g) *Thenkarakkera vellam*. This moves from south to east. Sea turns violent and encroaches upon the shore.
(h) *Vattakara neer*: This is from north to east and takes place in the month of *Chingam-Kanni* (September to November) having light winds.

Centres of Navigation on the Coromandel Coast

Mahabalipuram or Mammallapuram south of Madras was a flourishing port till its decline around the seventh century. Nagapattinam rose to importance once Mammallapuram gave way. The erection of a Buddhist *vihara* (called Chinese Pagoda) by Narasimhavarman Pallava for the benefit of Chinese merchants and of a Buddhist *stupa* built by Sailendra King Mara Vijayatungavarman sustained by a revenue endowment from the village Anaimangalam, granted by king Rajaraja in early eleventh century, are strongly suggestive of a prosperous maritime trade between the Cholas, South-East Asia and China.[57] There is no evidence to say that Nagapattinam was a centre of shipbuilding. It is concluded by scholars like Arunachalam that Vedaranyam or Topputurai further south was a centre of shipbuilding.[58]

The time of their navigation is this: from Persia to the Cape of Cumerin (Comorin), which is distant from Calicut eight days' journey by sea towards the south. You can navigate through eight months in the year, that is to say, September to April; then, from the first of May to the middle of August, it is necessary to avoid this coast because the sea is very stormy and tempestuous. And you must know that during the months of May,

June, July and August, it rains constantly night and day; it does not merely rain continually, but every night and every day it rains, and but little sun is seen during this time. During the other eight months it never rains. At the end of April they depart from the coast of Calicut, and pass the Cape of Cumerin, and enter into another course of navigation, which is safe during these four months, and go for small spices.[59]

Monsoons and Navigation in the Indian Ocean

Navigation in the Indian Ocean depended greatly on monsoon winds. The two monsoons namely the south-west monsoon and the north-east monsoon are well-known. Roman traders discovered the regularity of the monsoons in the Indian Ocean, in AD 47 attributed to a pilot named Hippalus. Roman ships began to sail directly to the port of Muziris (Muyirikottu) in Malabar. Egyptian Greek merchants who brought wine, brass, lead, glass, etc., for sale to Muziris and Bakare and purchased spices from there, sailed from Egypt in the month of July and arrived at Muziris in forty days. They stayed on the Malabar coast for three months and started their return journey from Muziris in December or January. Muziris was a centre of trade depending greatly on the monsoon. It is reported in the Sangam literature: 'The thriving town of Muziri, where the beautiful large ships of the Yavanas, bringing gold, come splashing the white foam on the waters of the Periyar which belongs to the Cherala, and return laden with pepper'.[60] It is further stated: 'Fish is bartered for paddy, which is brought in basket to the houses, sacks of pepper are brought from the houses to the market; the gold received from ships, in exchange for articles sold, is brought to shore in barges at Muziris, where the music of the surging sea never ceases, and there Kudduvan (the Chera king) presents to visitors the rare products of the seas and mountains'.[61] Strabo mentions that in his days he saw about 120 ships sailing from Myos Hormos to India.[62] The south-west monsoon generally lasts from May to August while the north-east monsoon spreads from November to February. The beginning of south-west monsoon on the Malabar coast usually takes place in the first week of June. It rains constantly during the months of May, June, July and August, night and day. Little sun was seen during these months. On the other hand during the rest of the year, i.e. for eight months of the year there was no rain on the Malabar Coast as testified by Ludovico di Varthema. Navigation in the Arabian sea in the vicinity of the Malabar coast was possible during these eight months from September to April. Since the sea was very stormy and tempestuous, navigation to and from the Malabar was avoided from the first of May to the middle of August. The ships from the Malabar coast departed by the end of April and passed the Cape of Comorin and entered into the other course of navigation into the Bay of Bengal and further. Navigation in this region for the subsequent four months was safe.[63]

Navigational Hazards

Even during the period of safe navigation, mariners had to be on their guard against the navigational hazards on the Malabar Coast. There are few indentations on this coast, though a number of estuarine mouths of large, medium and small rivers and a large number of tidal inlets provide breaks at almost regular, short intervals. A heavy surf prevails all along the coast on either side of Kanyakumari. An area of foul ground lies 2 km offshore, about 10 km west of Kanyakumari, 5 km south-west of Muttam (10 km west of Kanyakumari) lighthouse. There are several dangers, of which the Crocodile Rocks are the outer most, with a depth of barely 2 m. Others are the Adunda Rock and Kota Rock around which are all foul grounds. There are several rocky ledges around Colachel like Patna rock, Constance rock at depths of 2 m. There are other rocks such as Kurusukal, Ahanakal, Sudkal and Chadikal on the approach to Kolachel harbour. Shoals at 15 m depth are found off Kovalam.

Anjutenga port lying north of Tiruvananthapuram is exposed to surf which makes the approach to the coast very risky. Tangassery at the entrance of Ashtamudi Lake has about 2 km off shore, Gamaria Rock and Pallikkal shoal at 7 m depth. Extensive foul grounds with a depth of 1.2 m are situated off Nindakara. Shifting mud banks appear on surface especially during the south-west monsoon between Alapuzha and Kozhikode. The mud banks with oily surface provide safe anchoring grounds in the roadstead of Alapuzha and Kochi. The latter itself is backed by the backwaters leading to Vembanad and at its head is Wellingdon Island. There is a rock formation at a depth of 1.5 m about 1 km south of the Beypore River mouth. The mean level of water above this rock is 2.14 m. The river mouth has a bar, with a depth of 1.5 m. There is a Gillham rock with a depth of 1.8 m about 4 km south of Kozhikode lighthouse. Many detached shoals of 2 to 3 m are located to the north of the Coate reef at a depth of 5 m. The Reliance shoals at 8.6 m deep lie to the north-west of the lighthouse. A rocky islet is situated on them with an elevation of 1.8 m. There are further two rocky shoals at the depth of 5.9 and 6.4 m nearby. A bank of very soft mud with an easy landing is situated about 5 km north of Kozhikode.

A large black reef is found about 5 km north of Elattur River mouth besides a larger reef at 1.8 m depth. Kadalur reef a coastal reef at 5.5 m depth is situated to the west of Kadalur point called Kadalur reef with glistering white granite. Chombakkal islet is north of Vadakara with a height of 1.5 m. Talayi rock is found at a depth of 1.5 m on the approach of Tellicherry besides a foul ground with rocks above and below water nearby. Bilikalu, a natural breakwater of basalt lies parallel to the coast with numerous rocks above and below water level. Shorukulu ridge and Nakudiankulu are north of Tellicherry, but parallel to the coast with foul grounds. Green Island close to Dharmadam Island is fringed by a reef. The whole area is dangerous for navigation.

Kavvayi backwaters are situated adjoining Mt Dilli in the north. Bekal with a coastal fort and Kasaragod on the Chandragiri River are further north. Kumbala and Manjeswaram are small ports south of Mangalore.

Instruments of Astronomical Navigation

It can be asserted that towards the end of the fifteenth century or the advent of the Portuguese, India had a rich collection of the astronomical and time-measuring instruments either borrowed or developed indigenously.

Matsya Yantra

It consisted of an iron fish floating in a small vessel of oil. It pointed to the north. It was discovered in the eleventh century. It was the crude forerunner of modern magnetic compass. Sailors in Malabar used it for determining the direction.[64]

Mariner's Compass

Fully described in a Chinese work of the eleventh century, the mariner's compass must have reached Europe through the Arabs. It was used by the Europeans from the thirteenth century. The origin of modern magnetic compass may be attributed to the discovery of lodestone and its ability to attract iron. Though there is a passing reference to the use of this instrument in India in the work of Jacques de Vitry in his History of the Kingdom of Jerusalem (AD 1218) there is no conclusive information about it. It would appear that the Indian seamen might have used the magnetic needle, but they laid more stress on the determination of the direction by observing the position of the stars.

Stars

Stars and their constellation played an important role for the seamen in finding out the direction on board the vessels in the high seas. Every star has a fixed orbit in the sky, rising in the eastern horizon in the fixed azimuth[65] and setting likewise on corresponding fixed azimuth in the western sky. For the mariner, a rising or setting start was a very important indicator for azimuths or direction.

Druva Nakshatram

The sailors grouped the stars in constellations and gave names for the most bright ones. The pole star (*Druva Nakshatram*) is the clear star and is seen throughout the year except the months of *Kanni* and *Thulam* (October-November).

Kootunakshatram (Orion- crowded constellation)

The sailors paid great attention to this for the navigation in the sea. This is constellation of six stars in one line and ascending on the north-east of southern horizon at a time.

Kabar Nakshatram

Mapilla navigators of the Malabar coast term another star as *Kabar Nakshastram* (grave star) due to its position from south to north. It is a set of three stars that rise in the south-east.

Kibila Nakshatram is also a group of three stars rising in the north-east. These stars help one to find south and north and also to guide for sailing westward and eastward.

The star appearing first in this group is called *Mumbile Nakshatram*. This helps the seaman to steer the ship to the north. The stars are very useful in months of *Dhanu-Makaram (*November to January).

Another group of stars helpful in navigation is *Kappalar Nakshatram* – six ship stars. They are six in number. They are the main guiding stars for the navigators to reach north-east. If the sea-men sail slightly slanting, they would reach the northern side. If they steer in a direction opposite, the *Kappalar Nakshatram*, they will reach south-eastern direction.

The next group of stars is called *Thelinja Nakshatram*, or bright star. This consists of seven stars and is situated in the east to west direction. It sets around midnight. The brightness of this star is fully visible around 3 a.m. Other stars useful in navigation are *Ottanakshatram* (single star), *Erattanakshatram* (double star), *Nalunakshatram* (four stars) and *Kaodikkaran Nakshatram* (fisherman's star).

Measurement of Distance

The ancient Indian sailors had their own systems to measure distances, depths, and altitude. On the Malabar coast *viral kanakku* (finger calculation) was employed. It was the measurement of the width of a finger. The Arabs called it *isba* while the Portuguese navigators named it *polegada*. It was a unit of measurement of altitude. Twenty-four *virals* (fingers) made a cubit. One *viral* was equal to ¾ of an inch. The measurement of altitude using this is called *viralkanakku*. While using the *kamal* the *isba* or *viralkanakku* was employed.

Water clock of the sinking bowl variety was meant to measure time. It consists of a hemispherical copper bowl with a small perforation at the bottom. When this bowl is placed on the surface of water in a larger basin, the water enters the bowl from below through the perforation. As soon as the bowl is full, it sinks to the bottom of the basin with a clearly audible thud. The weight of the bowl and the size of the perforation are so adjusted that the bowl sinks sixty times in a day-and-night. That is to say, the duration between each two immersions is 24 minutes. It is believed to have been in India and neighbouring regions for about sixteen hundred years from the fourth century to the beginning of the twentieth century.[66]

Indian navigators had their indigenous equipment for measuring the altitude and finding the direction. When Vasco da Gama reached Melinde on the East African coast, the king of Melinde arranged a pilot for him to proceed to Calicut. He was Malemo Cana from Gujarat. When Vasco da Gama showed him a big wooden astrolabe and others made of metal to find out the altitude of the sun, the Muslim pilot did not show any sign of surprise. He said that in the Red Sea were used 'brass instruments of a triangular shape, and quadrants with which they took the altitude of the Sun and, especially, of the star [meaning the Pole Star] and which they used in navigation'. These were very useful for navigation. But he and the mariners of Gujarat as well as those of other parts of India counted on some stars of the north and south and other prominent ones which crossed through the sky by east or west. These mariners calculated the altitude through another instrument not similar to those, explained Malemo Cana. Then he showed the instrument to Vasco da Gama. This had three tablets.[67] This instrument as commented by Barros served the same purpose as those of the Portuguese. It was none other than the *kamal*.[68] This was like the cross-staff used in Europe. The Portuguese mariners on their return to Portugal under Vasco da Gama lost no time to try out *kamal*. They did not discard it immediately though it was difficult to use.

The mariners of the Coromandel coast had another contrivance to find out their latitudinal position off the coast.[69] This instrument consisted of a piece of thin board, oblong in shape, three inches long by one and half wide, furnished with a string suspended from its centre, eighteen inches long. A number of knots made in this string indicated certain previously observed latitudes, in other words coinciding with the positions of certain well known places on the coast.

Chinese Navigation in the Indian Ocean

Compasses had been employed by the Chinese to guide the directions for navigation at sea long before 2000 BC. After overthrowing the Mongol Dynasty in 1368, the Ming Dynasty came to power. Zhu Di, one of the sons of the first Ming emperor decided to re-open trade with foreign countries. Under his orders hundreds of ocean going vessels were built in the costal provinces. It is against this background that Admiral Zheng He started his voyages during the period from 1405 to 1433. His fleet travelled upto East Africa through the South China Sea, the Straits of Malacca, and across the Indian Ocean. In addition to the compass, Zheng He is believed to have used maps for navigation. There were detailed descriptions about what stars had been seen in the sky in different directions at certain locations at night. For example, in his map for the northern tip of Sumatra, it was illustrated that the Ursa Major's seven stars were at 'height' of one and a half 'fingers' above sea-level when facing north. (One finger equivalent to 1°36′.) Similarly, the Lyre was seen at certain angle facing the east, and some others were seen facing west and south respectively. Zheng He and his followers were able to sail across the Indian Ocean with these maps and compasses.[70]

The Kamal

A description of the *kamal* given by a Portuguese scholar seems to be quite clear.

> *Kamal* was made of one or more square or rectangular tablets through the centre of which was threaded a piece of string with knots at intervals from the tablet. To obtain the altitude of a star with this instrument, one chose the knot on the string which, held close to the eye of the observer or between the teeth, placed the tablet, when, held in front of the eye with the string taut, at such a distance that the pilot could sight the horizon along the lower edge of the tablet and the star at the upper edge[71]

The angle of the altitude measured in this way depended on the size of the board and the distance between the knots on the string.

But as the string was limited in length by the observer's reach, the *kamal* had two or even three boards, or else the board was rectangular in shape, so that the sight could be taken along either the two edges closer together or those further apart, according to convenience.

Here is still another description of the *kamal*.

It consists of a small parallelogram of horn (two inches by one) with a string or a couple of strings) inserted in the centre. On the string are nine knots. To use the instrument for taking the height of Polaris, the string is held between the teeth, with the horn at such distance from the eye, that while the lower edge seems to touch the oceanic horizon, the upper edge just meets the star: the division or knot is then read off as the required latitude.[72]

Prinsep describes the mode of marking off the knots.

Five times the length of the horn is first taken, as unit, and divided into twelve parts: then at the distance of six of these parts, from the horn, the first knot is made which is called '12'. Again the unit is divided into eleven parts, and six of these being measured on the string from the horn as before, the second knot is tied and denominated '11'. The unit is thus successively divided into 10, 9, 8, 7, and 6 parts, when the knot tied will of course exactly meet the original point of five diameters: this point is numbered '6' . Beyond it one diameter of the horn is laid off for the '5' division and one and a half again beyond, that for the '4' division which usually terminates the scale.[73]

The unit of measurement in the use of *kamal* is the *viral* (finger) and its subunits in eighths. It is equal to 0.75 of an inch. Twelve *virals* make a *saan* or *chan* and two *saans* a *muzham* (cubit), but different types of *muzhams* have been in use.[74]

The *kamal* described by Sidi al Celebi comprised nine small rectangular boards of varying small sizes with a string running through a hole at the centre of each board. Observations were made by holding one of the strings between the teeth and keeping one of the boards vertically at the other end of the string in such a manner that the lower edge was on the line of the visible horizon and the upper edge of the board parallel to the star under observation. Each board corresponded to a different elevation or altitude horizon, and the nine boards thus aid in measuring star elevations at different altitudes. The unit of measurement used is an *isba*.[75]

Vasco da Gama in his first voyage to India hired a pilot from the African coast to take him to the land of Calicut. It is not sure if the pilot was an Arab, Ibn Majid or a Gujarati Muslim, Malemo Cana.[76] However, it is generally accepted that the pilot used the *kamal.*[77] The Portuguese carried a copy of the same with them to Portugal. The tradition is now lost. Magnetic and stellar compasses were in vogue in the Arabian Sea and the Indian Ocean. The magnetic did not indicate one's present position though it indicated the direction. The altitude of Pole Star (*kavu*) unlike that of other stars, remains fixed throughout the night and through the year at a given location though it varies with geographical latitude. For short-distance travel, each increase in the altitude of the Pole Star is a measure of one's latitudinal position or of the north-south distance travelled. In using the *kamal*, the knots are counted by keeping the string between one's teeth. It is known with certainty that *kamal* was taken to Europe in 1499 by the sailors accompanying Vasco da Gama and it was used for a few years by the Portuguese navigators and cosmographers under the names *tavoletas da India* (Indian tablets) and 'Moorish cross-staff.'[78] The anonymous Florentine gentleman who accompanied Vasco da Gama makes two references to navigation in the Indian Ocean and to the use by is navigators of a 'certain wooden quadrant' (*certi quadranti di legno*).[79] It is generally accepted that these words of the Florentine man refer to *kamal* as affirmed by Luís de Albuquerque.[80] Master João who was charged with making observations with the *kamal*, the astrolabe, and the quadrant, during Cabral's voyage in 1500, confirmed that the *kamal* was the *tavoletas da India*, an expression found in Barros. In his letter to the king, Master João stated that he discharged the duty and made specific reference to the *tavoletas da India*.

From the oriental and Portuguese texts that contain direct, or indirect references to the subject, it appears that the *kamal* suited the navigation modes used between the coast of Arabia and the western and eastern coasts of India, and also the western coast of the Malaccan peninsula. *Kamal* was therefore used to obtain the altitudes of the Pole Star at the most frequented points on the opposite coasts, and these were marked by the navigators on the string of the instrument by a succession of knots. It could be possible that at first the *kamal* was constructed in this empirical fashion, each knot on the string marking the altitude of a star at two known points on the opposite coasts and on the same latitude, a practical procedure quite independent of any angular unit (*covado, isba* or any other). A passage copied into the *Livro de Marinharia* of André Pires shows the correct way of converting *isba* into degrees, i.e. 'if perchance you find a Moorish chart and wish to graduate it to our usage, you shall take 5 inches (*polegadas*) and divide them into 8 parts, which are eight degrees; thereby you shall obtain a chart graduated to our usage'.[81]

Rapalagai (Kavu-vellipalagai)

This was similar to the *kamal* and so some writers interchangeably use it with '*kamal*'. It was used for the determination of latitude at night. It literally meant night plate or a plate used at night. It is also known as *kauvellipalagai*. It was used to measure the altitude of stars at night and held by teeth. This was in vogue among the Tamil navigators on the Coromandel coast especially near Porto Novo. It is made of a rectangular board 3" × 2.25", i.e. 4 × 3 *virals*, with a hole in the exact centre through which runs a string knotted and fixed at the back of the board. The string itself has a number of knots. The knot positions correspond to specific ports of call determined and fixed with reference to the star used by repeated experience of visits. There exists a traditional *Rappalagai Sastram* among the boat-men of Lakshadweep islands. The *Rappalagai* of Lakshadweep consists of two rectangular ivory boards, with their length to width proportion being 3:2 while the Tamil *Rappalagai* or *kavuvelli palagai* had a proportion of 4:3.[82] *Rappalagai* helps in measuring the angles through holding the height constant and the distance varying. With it, the height of the board is constant, and the distance from the eye changes.

Speed

Speed was ascertained indigenously by comparing it with the pace of walking. The native sailor fully aware of the rate of his walking, 'throws a piece of wood overboard at the stern of the vessel and walks towards the stern keeping pace with the wood floating past, then he remembers his rate of walking, to which the progress of the vessel must necessarily be equal'.[83] This conversion of 8° into 5 *isba* makes the *isba* equal to 1° 36'.

The speed of the vessel was measured on the Malabar coast by tossing a wooden plank weighted with lead, known as *Tappu Palaka,* from the stern of the vessel and timing it using a sand glass. This plank was tied with a knotted thread and the length of the thread released during the period of sand fall in the sand glass gave the speed of the vessel. The system of calculation for conversion of the number of knots released over a time is called *tappukanakku*. It was the calculation of speed from the length of the thread released in a fixed time and was related to the length of the boat.

Indigenous Method of Docking a Ship

The ship was floated into a basin direct from the sea or inlet and the entrance was closed. The basin was surrounded by a high mud bank. The level of water in the basin upon which the vessel floated was raised by scraping the mud from the banks into the basin, levelling it at the bottom of the water, and so raising the bottom of the basin which would elevate the level of the water. This process was carried on until the ship was considerably higher than the level of the contiguous sea or inlet. The water was then suffered to run off, two beams were placed under the ship, stem and stern, resting on the new and exposed bottom of the basin. Perpendicular shores were then put to the ship and the earth levelled until the ship was on the same plain as the adjoining ground.

In order to undock, four sets of cables were used, each one was coiled into the shape of a solid cone, one fate or coil not touching the one beneath it, soft mud and sand being interposed between each layer as well as smeared all over it. One cone was placed under the starboard bulge forward, and another the same bulge aft, a third and a fourth corresponded in position on the larboard side; thus four solid cones of rope supported the ship. She was gradually lowered by withdrawing from the base of each cone simultaneously a coil or fake, by which the four cones bodily subsided and the vessel along with them, resting as she did upon them. It might be conjectured that by removing the lower coil the superstructure would tumble down, but this was obviated by the solidity of the mass, each layer consisting of a solid flat coil of rope one circle within another.[84]

Thus we find evidence of a rather well-developed art of shipbuilding and navigation in India. The Indian shipwrights and navigators could not continue to develop their art, however, on account of the arrival of the foreigners who imposed their own system of shipbuilding and navigation. The impact of this was felt in the period of industrialization.

NOTES AND REFERENCES

1. 'All these qualities [being tough, dry, of bitter and resinous sap, and pliable] are with difficulty found in one species of tree, and only Teak and *Angelim* have them, whose timbers appear incorruptible, and [it appears] that nature created them for Naval Architecture'. Ref. João Baptista Lavanha, *Livro Primeiro da Architecura Naval*, Lisboa, 1996, p. 141.
2. Kripa Shankar Shukla, 'Aryabhatta I's Astronomy with Midnight Day-reckoning', *Ganita*, June 1967, 18.1, pp. 83-4.
3. S.R. Sarma, 'Astronomical instruments in Brahmagupta's *Brahmasputa-siddhandta* '*Indian Historical Review*, 1986-7, pp. 63-74.
4. Iswara Candra Sastri, ed., *Yuktikalpataru Mahraja-Sri-Bhojaviracitah*, Calcutta, 1917.
5. Sreeramula Rajeswara Sarma, 'The Sources and Authorship of the Yuktikalpataru', *Aligarh Journal of Oriental Studies*, 1986, pp. 39-54.
6. S. Soundarapandian, ed., *Navai Sattiram* (Tamil), Madras, 1995.
7. B. Arunachalam, 'Timber Traditions in Indian Boat Technology', in K.S. Mathew, ed., *Shipbuilding and Navigation in the Indian Ocean Region AD 1400-1800,* Delhi, 1997, pp. 12-19.
8. T.P. Palaniyappa Pillai, ed., *Kappal Sattiram*, Madras, 1950.
9. The palm leaf manuscript was in the possession of R.Tirumalai in 1900 Ref. R. Tirumalai, 'A "Ship Song" of the Late 18th Century in Tamil', in K.S. Mathew, ed., *Studies in Maritime History*, Pondicherry, 1990, pp. 159-64.
10. Arunachalam, 'Timber Traditions in Indian Boat Technology', pp. 12-19.
11. Sreeramula Rajeswara Sarma, 'Instrumentation for Astronomy and Navigation in India at the Advent of the Portuguese', in Lotika Varadarajan, *Indo-Portuguese Encounters: Journeys in Science, Technology and Culture,* vol. II, Delhi, 2006, pp. 505-15.
12. B. Arunachalam, *Heritage of Indian Sea Navigation*, Mumbai, 2002, p. 5.
13. G.R.Tibbetts, *The Navigational Theory of the Arabs in the Fifteenth and Sixteenth Centuries*, Coimbra, 1969, p. 3, also ref. James Prinsep, 'Extracts

from the *Mohit,* that is the Ocean', Turkish work on Navigation in the Indian Seas, tr. J. Hammer-Purgstall, *Journal of the Asiatic Society of Bengal*, 1834, pp. 545-53; 1836, pp. 441-68; 1837, pp. 805-12; 1838, pp. 767-80; 1839, pp. 823-30, Calcutta, 1834-9.

14. Arunachalam, *Heritage of Indian Sea Navigation*, p. 6.
15. *Ibid.*, p. 8.
16. C.M.S. Chandera, *Kannakiyum Cheermmaakkavum*, Kottayam, 1973.
17. K.K.N. Kurup, 'Indigenous Navigation and Shipbuilding on the Malabar Coast', in K.S. Mathew, ed., *Shipbuilding and Navigation in the Indian Ocean Region AD 1440-1800*, Delhi,1997, pp. 20-5.
18. B. Arunachalam, *Chola Navigation Package*, Mumbai, 2002, p. 34.
19. Fernando Oliveira, *O Livro da Fabrica das Naus,* Lisboa, 1991, pp. 146-50.
20. João Baptista Lavanha, *Livro Primeiro da Architectura Naval*, Lisboa, 1996, pp. 143-6.
21. V. Kunhali, 'Timber Industry Related to Shipbuilding in Kerala', in G. Victor Rajamanickam and Y. Subbarayalu, eds, *History of Traditional Navigation*,Thanjavur, 1988, pp. 159-60.
22. The word '*mestiri*' takes it origin from the Portuguese word *mestre*, meaning master.
23. Some ships are of flat bottom, others have V bottom and still others round bottom.
24. A.P. Greeshmalatha and G. Victor Rajamanickam, 'The Ship-building Technology as Practised in Beypore, Kerala', in K.S. Mathew, ed., *Shipbuilding and Navigation in the Indian Ocean Region, AD 1400-1800,* Delhi, 1997, p. 50.
25. Arunachalam, 'Timber Traditions in Indian Boat Technology', p. 15.
26. For details on sewn boats, ref. Eric Kentley, 'The Sewn Boats of India's East Coast', in Himanshu Prabha Ray and Jean-François Salles, eds, *Tradition and Archaeology: Early Maritime Contacts in the Indian Ocean*, Delhi, 1996, pp. 247-60.
27. Ludovico di Varthema, *The Itinerary of Ludovico di Varthema of Bologna from 1502 to 1508*, London, 1928.
28. Radhakumud Mookerji, *Indian Shipping: A History of the Sea-borne Trade and Maritime Activity of the Indians from the Earliest Times*, Delhi, 1999, p. 41. The representations of ships are found in Cave 2, of Ajanta which are placed between AD 525 and 650.
29. Mookerji, *ibid.*, p. 50.
30. *Ibid.*, pp. 51-2.
31. Varthema, *The Itinerary of Ludovico di Varthema of Bologna from 1502 to 1508*, p. 62.
32. *Ibid.*, p. 62.
33. Adelino de Almeida Calado, *Livro que trata das cousas da India e do Japão:*

edição critica do codice Quinhentista 5/381da Biblioteca Municipal de Elvas, Coimbra, 1957, pp. 67-8.

34. Oliveira, *O Livro da Fabrica das Naus*, p. 153.
35. Calado, *Livro que trata das cousas da India e do Japão*, p. 69.
36. Oliveira, *Livro da Fabrica das Naos*, p. 153.
37. K.N. Chaudhuri, *Trade and Civilisation in the Indian Ocean: An Economic History from the Rise of Islam to 1750*, Delhi, 1985, pp. 151-2.
38. M.R.R. Warrier and Rajan Gurukkal, *Kerala Charithram*, Edappal, 1991, p. 189.
39. Mookerji, *Indian Shipping*, pp. 103-4.
40. Varthema, *op. cit.*, p. 62.
41. Barbosa, *op. cit.*, vol. 2, p. 76.
42. *Ibid.*, p. 76.
43. Varthema, *op. cit.*, p. 62.
44. João de Barros, *Da Asia*, Decada I, Lisboa, 1777, p. 425; Castanheda speaks of a ship belonging to a merchant of Cochin called Patemarakkar which carried an elephant and three hundred men of arms among other things. The ship was captured by the men of Pedro Álvares Cabral as desired by the Zamorin in 1500. Ref. Fernão Lopes de Castanheda, *Historia do Descobrimento e Conquista da India pelos Portugueses*, Livro I, Coimbra, 1924, pp. 83-5. Gaspar Correa on the other hand refers to a ship belonging to a merchant of Cochin which carried an elephant and other commodities to Gujarat via Calicut. This was a huge ship . It was captured by the men of Álvares Cabral as the Zamorin wanted to get the elephant carried on board the ship. Ref. Gaspar Correa, *Lendas da India*, tomo I, Coimbra, 1922, pp. 196-203.
45. João de Barros, *Da Asia*, Decada I, part 2, Lisboa, 1777, pp. 29-38.
46. Varthema, *op. cit.*, pp. 62-3.
47. *Ibid.*
48. Arunachalam, *Chola Navigation Package*, 2004, pp. 36-7.
49. *Dhawar* is an Arabic term meaning 'go around'.
50. Varthema, *op. cit.*, p. 62.
51. Captain H. Congreve, 'A brief notice of some contrivances practised by the native mariners of the Coromandel Coast in navigating, sailing and repairing their vessels', in Gabriel Ferrand, ed. *Introduction a l'Astronomie Nautique Arabe*, Paris, 1928, pp. 28-9.
52. Diogo Köpke, ed., *Primeiro roteiro da costa da India; desde Goa até Dio: narrando a viagem que fez o Vice-Rei. D. Garcia de Noronha em soccoro desta ultima cidade, 1538-1539 por Dom João de Castro, Govenador e Vice-rei , que depois foi, da India*, Porto, 1843, pp. 173, 189, 192-4, 196, 197.
53. Sadashiv Gorakshkar and Kalpana Desai, *The Maritime Heritage of India*, Bombay, 1989, p. 8.
54. Genesis, Chapter 8, verses 6-8.

55. Pliny VI, 22, ref. Radha Kumud Mookerji, *Indian Shipping*, p. 103
56. Arunachalam, *Chola Navigation Package*, 2004, pp. 64-7.
57. *Ibid.*, p. 29.
58. *Ibid.*, p. 32.
59. Varthema, *ibid.*, p. 62
60. *Erukkaddur Thayan Kannanar - Akam,* 148.
61. *Oaranar-Puram*, 343.
62. Strabo, ii, v. 12.
63. Varthema, *op. cit.*, p. 62.
64. K.M. Panikkar, *India and the Indian Ocean*, Bombay, 1971, p. 26.
65. Azimuth is an angle related to distance around the earth's horizon, used to find out the position of a star, planet, etc.
66. Sarma, 'Instrumentation for Astronomy and Navigation in India at the advent of the Portuguese', *op. cit.*, pp. 505-15.
67. João De Barros, *op. cit.*, Decada I, part I, pp. 319-20.
68. Luís de Albuquerque, *Instruments of Navigation*, Lisboa, 1988, p. 29.
69. Ref. Captain H. Congreve, Madras Artillery; 'A Brief Notice of Some Contrivances Practised by the Native Mariners of the Coromandel Coast in Navigating , Sailing and Repairing their Vessels', in Gabriel Ferrand, *Introduction a l'Astronomie Nautique Arabe,* Paris, 1928, pp. 25-30; and in *Madras Journal of Literature and Science,* vol. XVI, January-June 1850, pp. 101-4 (ed. Madras Literary Society.)
70. Walter Lenz, 'Voyages of Admiral Zheng He before Columbus', in K.S. Mathew, ed., *Shipbuilding and Navigation in the Indian Ocean Regions AD 1400-1800*, Delhi, 1997, pp. 147-54.
71. Albuquerque, *Instruments of Navigation*, p. 30.
72. James Prinsep, 'Note on the Nautical Instruments of the Arabs', in Gabriel Ferrand, *Introduction a l'Astronomie Nautique Arabe,* Paris, 1928, pp. 1-23; and in *Journal of the Asiatic Society of Bengal*, vol. 5, December 1836, pp. 784-98.
73. Prinsep, 'Note on the Nautical Instruments of the Arabs', *ibid.*, pp. 2-3.
74. Arunachalam, 'Indigenous Traditions of Indian Navigation', p. 136.
75. Arunachalam, *Heritage of Indian Sea Navigation*, p. 44.
76. João de Barros, *Da Asia*, Decada I, part I, p. 319.
77. *Kamal* means complete. It is also known as *rapalagai* meaning block of wood or instrument used at night, C.K. Raju, 'Kamal or Rapalagai', in Lotika Varadarajan, ed., *Indo-Portuguese Encounters: Journeys in Science, Technology and Culture*, vol. II, New Delhi, 2006, p. 483.
78. Albuquerque, *Instruments of Navigation*, p. 29; also G.R.Tibbetts, *The Navigational Theory of the Arabs in the Fifteenth and Sixteenth Centuries*, Coimbra, 1969; José Manuel Malhão Pereira, *East and West Encounter at Sea*, Lisboa, 2002; José Manuel Malhão Pereira, *The Stellar Compass and the Kamal,* Lisboa, 2003.

79. Ramusio, Primo volume & seconda editione *Delle navigationi et viaggi*, Venice, 1554, '. . . I marinari di la [Calicut] cioé i Mori non nauigano con la tramontana, ma con certi quadranti di legno' 131.C; 'Et appresso afferma, che nauigano in quelli mari [the Persian Gulf] senza bussolo, ma co certi quandranti di legno, che pare dificil cosa, et massime que fa nuuolo , che no possono vedere la stele' 128 B. With reference to the latter quotation, Gabriel Ferrand , Introduction à la'astonomie nautifque arabe, Paris, 1928, p. 168 explains how the navigators of the Indian Ocean made observations of altitude even when the Polar Star could be seen.
80. Albuquerque, *Instruments of Navigation*, p. 29; alsoTibbetts, *The Navigational Theory*, p. 29.
81. Luís de Albuquerque, *Livro de Marinharia de André Pires*, Lisboa, 1963, p. 220: 'Se caso for que achares alguma carta de mouros e a quiseres graduar à nossa usança, tomarás 5 plegadas e reparti-las-ás em 8 partes, que são 8 graus; e por ali tirarás uma carta graduada á nossa usança'.
82. For the details of its use, ref. Arunachalam, *Heritage of Indian Sea Navigation*, pp. 48-52.
83. Captain H. Congreve, 'A Brief Notice of Some Contrivances Practised by the Native Mariners of the Coromandel Coast in Navigating, Sailing and Repairing their Vessels', in Gabriel Ferrand, *Introduction a l'Astronomie Nautique Arabe,* Paris, 1928, p. 28.
84. *Ibid.*, pp. 29-30.

CHAPTER III

The Arrival of Portuguese in India: Early Interactions

With a view to understanding the maritime activities of the Portuguese on the western coast of India and the interactions of Indians with the Portuguese, a short description of the arrival of the Portuguese and their settlements in India will be given below. The Portuguese entered into friendly relations with a few native rulers and obtained permission to set up their factories and fortresses in strategically important places on the west coast such as Diu, Daman, Bassein, Dabol, Chaul, Goa, Cannanore, Calicut, Cranganore, Cochin and Quilon. They set up shipbuilding centres attached to these establishments. They looked forward to gaining access to the interior from where various varieties of timber suited to the building of ships could be brought. After a long voyage from Lisbon to India they needed installations for careening the ships, and even building new ships. They always took into account the availability of suitable timber when they set up installations for shipbuilding.

Attempts at the Discovery of Direct Sea-route

The discovery of the oceanic routes to the West and East Indies has been considered 'the greatest event since the creation of the world, apart from the incarnation and the death of Him who created it' by Francisco Lópes de Gómara in his work *General History of the Indies* in 1552. Almost in the same vein, Adam Smith viewed the discovery of America and the passage to the East Indies via the Cape of Good Hope as the two greatest and most important events in the recorded history of mankind.[1]

Early Attempts

By Sea

Two ships of 50 tonnes were made ready for the voyage of Bartholomeu Dias who left Portugal in August 1486. Pero d'Alanquer was the captain of the ship under Bartholomeu Dias, while João Infante was the captain of the second ship. A third ship carried the provisions under Pero Dias, brother of Bartholomeu Dias as the captain.[2] The ships rounded the Cape of Good Hope early in 1488 and proceeded some distance up the coast of southern Africa. Dias returned home with the message that the sea route to the Indies was open.

By Land

Dom João II (1481-96) king of Portugal decided to discover the East through land route via Jerusalem, and sent Frei Antonio de Lisboa and Pero de Montaroyo to Jerusalem. But the results were not satisfactory since his men did not know Arabic and did not dare proceed further. Knowing the importance of Arabic language for this endeavour, the king ordered Pero Covilhã, an Arabic speaking person, and his companion Afonso de Paiva to take up this work. They started from Santarém on 7 May 1487[3] and went to Naples from where they proceeded to the island of Rhodes and then to Alexandria. After a stay there shortened by illness, they made their way to Cairo and Tor in the company of some Muslims going to Aden. Afonso de Paiva took his route to Ethiopia, while Pero Covilhã left for India. Covilhão set sail in a small vessel to Aden and from there visited Cannanore, Calicut, and Goa. He returned to Mina de Soffala (Ethiopia). Again he went to Aden and finally to Cairo where he got the information that Paiva had died of some illness. There he came in contact with two Jews from Spain with whom two Jews from Portugal, Rabi Habrã hailing from Beja and Josepe the shoemaker from Lemego, were on the look out for Pero Covilhã. The latter knowing that the king of Portugal was eager for information regarding India, had collected interesting information about Babylonia which was known as Baghdad situated on the Euphrates, and Ormuz in the Persian Gulf where a flourishing trade in spices and the riches of India was going on. He had also information about the caravans carrying commodities to Damascus and Aleppo. By the time the king obtained these details from the Jew, Pero Covilhã had already departed from Santarém. So he had instructed Josepe and Rabi Habrão to go in search of Covilhã and give him the letter from the king. Rabi Habrão was asked to go with Covilhã to Ormuz and from there to collect information about India. So he proceeded to Aden from where both of them set sail for Ormuz to gain knowledge about caravans going to Aleppo. Covilhã returned to the Red Sea and finally to the Court of Prester John after sending a detailed report of all his findings to the king from Cairo in 1490/1.[4] It is not certain if this report ever reached the king.

The Voyage of Vasco da Gama

When João II, king of Portugal died without any legal heir to succeed him, the Duke of Béja, Dom Manuel, his cousin, took possession of the kingdom on 27 October 1495 at the age of twenty-six, according to the testament of king João II.[5] Dom Manuel (1496-1521) in the second year of his reign sent the first fleet to India under the leadership of Vasco da Gama who was picked up by D. João II, the king. The king in the course of his discourse before sending the fleet referred to the power wielded and the riches accumulated by Venice, Genova, Florence and other Italian centres through trade and commerce. He enthused the Portuguese about maritime trade with India.[6] He entrusted to Vasco da Gama the banner of the Order of Christ and a letter to be handed over to the Zamorin of Calicut. Vasco da Gama left for India on 8 July 1497 with four vessels, namely *São Gabriel* commanded by Vasco da Gama, *São Rafael* under Paulo da Gama his brother, and *Bérrio* led by Nicolau Coelho, and a fourth that carried the provisions for the mariners under captain Gonçalo Nunes.[7] The average tonnage of these vessels was around 100 to 120. There were approximately 160 persons on the voyage. Vasco da Gama while in Melinde on the East African coast obtained the services of a Gujarati pilot who took the fleet to Kappad about 12 km north of Calicut. It is believed that the fleet anchored off Kappad on 18 May 1498. On the next day in the light of the information received from a fisherman the fleet proceeded to Calicut. A Portuguese *degredado* landed at Calicut on 20 May 1498[8] and passed a message to the Zamorin of Calicut, who at that time was in Ponnani. The latter sent a pilot to direct the fleet to Pantalayani, situated about 24 km north of Calicut where the Portuguese reached on 27 May 1498.[9]

Pantalayani Kollam along with Calicut had attracted the attention of the Chinese traders from the thirteenth century. Mention is made of Chinese trade with Pantalayani or Fandaraina in a document of 1296.[10] Ibn Battuta, hailing from Tangier, on his way to China had visited the Malabar coast in 1342. Fanderayana (Pantalayani) figures among the port towns mentioned by him. The other towns he is reported to have visited were Hili (Mount Deli), Jurfattam, Dahfattam and Calicut. He saw about 13 Chinese ships at Calicut.[11] Since Calicut did not have suitable anchorage during the monsoon ships are presumed to have anchored at Pantalayani which has a sheltered anchorage. Ibn Battuta writes:

> Then we left Budafattan for Panderani (Fandarayna) – a large and beautiful city with gardens and bazaars. There are three Muslim quarters each of which has a mosque, while the congregational mosque lies on the coast. It is wonderful, and has observation-galleries and halls overlooking the sea. The judge (Qazi) and the orator (*Khatib*) of the city is a man from Oman, and he has a brother who is accomplished. It is in this town that ships from China winter.[12]

The information regarding the presence of Chinese junks makes it clear that the port at Pantalayani was able to provide safe anchorage for ships with great tonnage. This also gives the indication that Calicut which had only an open sea was not suitable during the monsoon for the anchorage of ships for a long time and so they used to be anchored in the port of Pantalayani Kollam.

This is confirmed by the steps taken by the Portuguese at the suggestion of the Zamorin of Calicut. The first fleet under Vasco da Gama reached Calicut in May 1498 just before the beginning of the monsoon in June. The Zamorin who was at that time in Ponnani received the message regarding the arrival of the Portuguese at Calicut.[13] Immediately the Zamorin sent a pilot with the two messengers with orders from the Zamorin to take them to Panderani where the anchorage was better and safer.[14]

The presence of a Qazi from Oman could easily lead us to the conclusion that there were a number of Omanese traders settled in Pantalayani along with other foreign merchants. He could be like the *Shabandar* who looked after the well-being of the merchants in foreign land. There was always a *Shabandar* in Malacca to look after the affairs of Gujarati merchants settled in Malacca as well as those who frequented the port regularly from Gujarat.

Pantalayani was considered a port town of importance in the first decade of the sixteenth century as reported by Duarte Barbosa, a Portuguese official who worked in the Portuguese factory at Cannanore from 1503 and later at Calicut when the Portuguese were able to set up a fortress in 1513.[15] Writing in 1515, Tomé Pires speaks of Kollam otherwise known as Pantalayani (Pamdaranj) as one of the ports on the Malabar coast providing anchor to ships. He adds that the port of Pantalayani belonged to the Zamorin.[16]

Accompanied by ten of his companions Vasco da Gama set out on 28 May to call on the Zamorin. On his way he visited a Hindu temple at Puthur dedicated to goddess Durga, and another temple. The contacts with the people of Calicut were established through Fernão Martins with Vasco da Gama, the only person who spoke Arabic. The letter sent by King Manuel to the Zamorin of Calicut was in Arabic with translation in Portuguese.[17]

The Return of Vasco da Gama

According to the information provided in the journal of the first voyage of Vasco da Gama, the Portuguese fleet spent 101 days at Pantalayani during the monsoon season. This was possible only because of the safety of the port. The fleet left Pantalayani on 29 August 1498 for Portugal with a letter from the Zamorin to King Manuel.[18] The fleet reached Lisbon on 10 July 1499.[19] On account of the success of the fleet in discovering a direct sea-route to the Malabar coast, the king assumed the title '*Senhor da Conquista, Navegação, e commercio da Ethiopia, Arabia, Persia e India*' and rewarded Vasco Da Gama appropriately, also granting the title 'Almirante dos mares da India'.[20] Infant Dom Henrique had already built a house in honour of Blessed Virgin Mary at Restello under the invocation of Bethlehem in which a few members of the military order of Christ used to stay. He was the Governor and Administrator of the Order of Christ. A lot of immovable property around it was donated by him with the condition that the Chaplain should offer a weekly mass in honour of Blessed Virgin Mary for Dom Henrique. The priests residing there were expected to look after the administration of sacraments like Confession and Eucharist to those who used to take up overseas activities. Dom Manuel who succeeded Dom Henrique as the Governor and Administrator of the Order of Christ, as soon as Vasco da Gama returned to Lisbon from India wanted to build a huge church at the hermitage known under the invocation of Bethlehem. He built a magnificent church in a more prominent site and entrusted it to the Religious of the Order of St. Jerome for the glory of Blessed Virgin Mary. He opted for the same place for his tomb.[21]

Information about the successful endeavour of Vasco da Gama was passed on to all the Portuguese towns and villages by the king asking residents to thank God for the graces showered on the nation in connection with the marvellous victory of Vasco da Gama in reaching the Indian shores via Cape of Good Hope. When the Portuguese people saw the pepper, cloves, cinnamon, seed pearls and, precious stones brought by the fleet of Vasco da Gama which were till now taken to Lisbon on Venetian Galleys they praised the king for his initiative and stated that Dom Manuel was the most fortunate of the Portuguese kings who within two years of his coronation accomplished marvellous deeds. The Portuguese concluded from the experience of Vasco da Gama that the best time to start for India was March.[22]

After realizing the importance of the maritime endeavour, a decision was taken by the king's council to send a large fleet under Pedro Álvares Cabral which would be able to give an idea about the greatness of the kingdom. The date of departure was fixed as 8 March 1500. The fleet consisted of thirteen vessels of different types such as *náos, navios* and caravels. Eight monks of the Franciscan order under the leadership of Fr. Henrique and a vicar who was appointed to administer sacraments in the proposed factory to be built in Calicut, accompanied the fleet.[23] Cabral reached Calicut on 13 September 1500 after having made a diversion to Brazil. Gaspar da India functioned as an interpreter in the meeting with the Zamorin. The letter written in Arabic by the king of Portugal to the Zamorin made mention of the intention of the visit to deal with details related to peace and friendship with the Zamorin and trade in spices and other matters. The Portuguese king offered whatever was needed for the defence of his kingdom in terms of personnel and arms and ammunition. The letter further noted that gold, silver and other merchandise for the purchase of spices were brought to Calicut under the leadership of the captain Major Pedro Álvares Cabral. Since the Captain knew that the Zamorin was friendly with his neighbours he was to take care only of trade and commerce with Portugal. It was against this background that the Portuguese requested the permission to open a factory in Calicut. A treaty of peace and friendship between the Zamorin and the Portuguese was concluded at a place near the seashore (*Cerame* or *Srambi*).

After several discussions, the treaty was concluded and a factory near the sea was opened where the Portuguese flag was hoisted. Fr. Henrique, though he was not fluent in Malayalam, was put in charge of the religious activities. The Zamorin signed two copies of the treaty, one on a silver plate with a gold seal and the other on a copper plate with a brass seal. The former was to be taken to the Portuguese king and the latter to be kept in the factory at Calicut.[24] In view of the commercial treaty, a factory near the seashore with Aires Correa as its official was set up in a building given by the Zamorin and Aires Correa was appointed as the factor. He was given a Gujarati merchant to instruct him in the customs and manners of the country.[25]

But there arose problems in obtaining the required cargos of spices on account of the alleged interference of the merchants dealing with Mocha. Having managed to get only old pepper for two vessels, the Portuguese were looking for more cargo. In the meantime, a ship belonging to Mammale Marakkar and Cherina Marakkar of Cochin, loaded with seven elephants from Ceylon was passing through the Arabian Sea to Mocha. Zamorin showed great interest in the elephant and wanted the Portuguese to get hold of the ship for his sake which they did to please the Zamorin as advised by Coge Cemetery, a powerful merchant of Calicut. In fact the Portuguese were told that the ship belonged to merchants from Mocha and that it was carrying pepper, which was not true. The Muslims of Calicut turned against the Portuguese on account of the capture of the ship. Further the Portuguese, frustrated in their effort to get more cargo, searched a ship loaded with pepper for Mocha. In the ensuing fight the indignant Muslims murdered Aires Correa the factor and about fifty men with him in the factory. Thus the first factory established on the Malabar coast was razed to the ground.[26]

Apprised of the sad event, Vasco da Gama came in 1502 for the second time to Calicut with a fleet of twenty-five vessels to take revenge upon Calicut. When diplomacy failed he stormed the city and massacred a lot of people. He sent in a boat heads, arms and legs separated from their trunks as a 'present' to the Zamorin with a letter written in the local language saying that he came to Calicut to sell and purchase good commodities and these were the merchandise he could find there.[27] After this encounter, the Zamorin started building up technical know-how to fight the Portuguese. At the request of the Zamorin the sultan of Cairo sent four Venetians, experts in artillery, to Calicut among whom was, presumably, Ludovico di Varthema.[28]

The port of Calicut was considered by Afonso de Albuquerque, the Portuguese governor the best in India and its ruler the most powerful of the princes of Malabar. It had an abundance of commodities and rich merchants.[29] Soon after the demise of the reigning Zamorin, his nephew / his younger[30] brother sent a message in 1512 to Afonso de Albuquerque in Goa expressing his willingness to come to terms with the Portuguese and permitting them to have a factory and a fortress in Calicut. Afonso de Albuquerque had already written to the king of Portugal explaining the importance of the port of Calicut and the power of the Zamorin of Calicut. He wrote very emphatically that the Zamorin was the most powerful of all the rulers of Malabar and that the port of Calicut was the biggest of all ports in India and the trade in this port was great. Very rich and powerful merchants were found in Calicut.[31] He had informed the Portuguese king that Calicut was the ancient emporium of trade with Cairo and Venice.[32]

Afonso de Albuquerque sent his nephew Garcia de Noronha to arrange the treaty of peace with the king of Calicut. He posted Francisco Nogueira as captain of the proposed fortress, Gonçalo Mendes as the factor, and Thomas Fernandez as the master of masons to build the factory.[33] He underlined the fact that Calicut would be a true emporium for pepper and ginger for Portugal and would provide cargo for ships coming from Cochin.[34] He further reasoned out that the only way to stop the flow of spices from Calicut to Cairo was to have a fortress in the place offered by the Zamorin and to have eighty men in the fortress. It would be rather impossible to stop the diversion of spices through waging maritime battles.[35] Garcia de Noronha on behalf of Afonso de Albuquerque, the Portuguese governor made the agreement with the Zamorin of Calicut on 24 December 1513.[36] The Portuguese took the responsibility of bringing to Calicut coral, silk cloths, scarlet, quicksilver, vermilion, lead, copper, saffron, alum stone and other commodities available in Portugal. The Zamorin undertook to supply all the necessary spices and drugs found in the region of Malabar and needed for the Portuguese.

The Portuguese bound themselves to pay the usual taxes for the commodities they purchased. The buyers of the commodities were asked to pay customs duties to the king of Calicut. The merchants from Ormuz, Cambay, Malacca, Sumatra, Pegu, Tennasserim, Bengal, Coromandel, Ceylon, Jafnapattanam, and Chael were for their part bound to pay taxes to the king of Portugal. Similarly, the Portuguese who brought horses or elephants to Calicut were asked to pay tax to the Zamorin. Sambuks coming to Calicut were expected to collect *cartazes* from the Portuguese captain at Calicut. If any damage were caused to Portuguese interests in Calicut, the Zamorin agreed to deliver one thousand *bhares* of pepper in three instalments as compensation according to the unit of weight used in Cranganore. Besides, the local people were to be punished by the Zamorin and the Portuguese by their captain for any crime committed by them. The Zamorin agreed to pay to the Portuguese half of the income from the customs houses at Calicut. As desired by the Zamorin, the treaty was ratified by the king of Portugal on 26 February 1515.[37] The Zamorin further sent a relative to Portugal. The boy sent by Zamorin became Christian and received the habit of the Order of Christ. He was called Dom John of the Cross and lived in Portugal for five years.[38] Fifty Nairs worked as guards in the fort of Calicut in 1514 and out of them 20 used to reside in the fort night and day. They were under a Panikkar. All of them were paid a special amount and materials during the local festivals.[39]

But for the fact that the new Zamorin (1513-22) saw that his best interests lay in peace, not in war, this treaty would not have survived the death of Albuquerque in 1515. Lopo Soares, for example, demanded that the Zamorin should repair the Portuguese fort and wait upon him and hostilities were averted only by the good sense of the Portuguese captains, who refused to draw their swords in such a silly cause.[40] The king of Cochin did not like the treaty and so he looked for opportunities to create hostilities between the Portuguese and the Zamorin. The Portuguese too insisted on passes for the Muslim merchants once they completed the building of the fortress. The death of the Zamorin who concluded the treaty brought the matters to a head. The new Zamorin (1522-31) was less friendly than his predecessor. In 1523 the Moors insulted the Portuguese governor Duarte Meneses,[41] and in 1524 an open fight took place in the bazaar between them and some Portuguese soldiers.[42] The captain of the fortress submitted exaggerated reports about the event which precipitated the crisis.[43]At this juncture Vasco da Gama arrived in Goa as the Viceroy. In September he sent D'Souza with 300 men to assist the captain at Calicut. Vasco da Gama on reaching Cochin adopted more vigorous measures against the Zamorin. He died on 24 December and Henry Meneses succeeded him. There followed fights between the Portuguese and the people of Calicut. The fortress was as strong as the fortress of Cochin and in the same shape, but it was abandoned by the Portuguese in 1525[44] because of constant disturbances from the local merchants.

Fortress at Chaliyam

A new Zamorin came to the throne in 1531. The Portuguese were interested in building a fortress at least in the neighbourhood of Calicut to stop the flow of a great deal of pepper and other spices to the Red Sea regions. It is believed that the governor gave some gifts to extract consent of the ruler of Chaliyam.[45]

The Rajas of Bettet, Beypore and Chaliyam who had been vassals of the Zamorin gave up their allegiance to the Zamorin. The Portuguese tried to build a fortress at Tanur but for some reasons did not succeed. Then they turned to Chaliyam on the road to Ponnani and Cochin. Its Raja, Unni Rama heard them out but did not want to irritate the Zamorin. The Portuguese thus sent a messenger to the Zamorin seeking his approval. The Zamorin approved of it and accepted the offer of half of the customs duties on the traffic that passed through the river, and the Portuguese fortress at Chaliyam in 1531.[46] A church, a house for the captain, the soldiers and an arsenal were constructed at Chaliyam under the orders of the governor, Nuno da Cunha. The ruler of Chaliyam was approached to have *jangada* (*changathi*) for the security of the fortress.[47] The Chaliyam River was very suited for the anchorage of large number of vessels and also for cargos of pepper. It was considered to be the best for navigation on the Malabar coast.[48]

However, peace with the Zamorin lasted only for a few years.[49] The Zamorin sent Chinakuttiali, a merchant of Calicut in 1539 to the Viceroy in Goa asking for peace and friendship. At that time, the Portuguese fortress at Chaliyam was under the Captain Manuel de Brito.[50] The ruler of Tanur represented to the king of Calicut while the details of the terms of the agreement were finalized. The Portuguese Viceroy Dom Garcia de Noronha made a treaty with the king of Calicut on the ship *St. Mattheus* at Ponnani on 1 January 1540.[51] Accordingly the king of Calicut promised to supply necessary pepper and other commodities to the Portuguese, to stop sending ships to Mocha and receive ships from there. The peace lasted for ten years. Fights between the Portuguese and the Zamorin broke out in 1550 and D. Francisco Coutinho, appointed Viceroy in 1561, established peace with the Zamorin only in 1662.[52] He started from Cochin and met the Zamorin near Ticodi. The Zamorin was accompanied by a militia consisting of 40,000 Nairs and a number of Brahmins, Kaimals, and Panikkars and so on. The viceroy too had a big contingent. After usual courtesies, the Zamorin made an oath assisted by his Brahmin priests and the Viceroy took the oath touching the Missal and the Crucifix.[53]

The fortress at Chaliyam was destroyed by the Zamorin in 1570. D. Jorge de Castro was the captain of the fortress of Chale when it was besieged by the Zamorin.[54] Ultimately the Portuguese left Chaliyam with their wives on 4 November 1571.[55] The fortress was completely razed to the ground.

Fortress at Ponnani

Ponnani was an important centre of trade. It was between Calicut and Cochin. There were several merchants who traded with Mocha and via Red Sea with other parts of the world.[56] In view of the Portuguese attempts to stop the flow of commodities to the Red Sea regions, the merchants of Ponnani were totally unhappy with the Portuguese. In November 1507 itself Kutiali, a powerful captain of the Zamorin having with him more than seven thousand armed men, battled with the Portuguese in November. Tristão da Cunha defeated them and seized their arms and ammunitions. The commodities that were made ready for despatch to Mocha were also taken over by the Portuguese.[57]

However the Zamorin in 1584 permitted the Portuguese to establish a fortress at Ponnani. The Portuguese viceroy, D. Duarte de Meneses being aware of the importance of the river at Ponnani and the necessity of stopping the flow of spices to the Red Sea regions and also to divert the attention of the Turks by sending a fleet from here to the Red Sea, took up the matter of setting up fortress at Ponnani.[58] Dom Jeronimo Mascarenhas was instrumental in signing the contract with the Zamorin.[59] So he was appointed the captain of the fortress at Ponnani and Ruy Gonsalves de Camara was nominated captain of the north and of the Armada in 1585.[60] Rui Gomes de Gram in the capacity of captain of the fortress strengthened the defences in course of time.[61] Later Rui Gomes, the captain of the fortress visited the Zamorin at his residence and paid homage.[62] He was given a rousing welcome in the presence of Managat Achan, the chief of the administrators of the Zamorin and others.

By 1586 the relations between the Zamorin and Kunhali were not good. The Gaspar Fagundes who was in the fortress at Ponnani was asked by the Governor to offer his services to the Zamorin against Kunhali.[63] A few Portuguese were captured by Kunhali and were insulted later in his fortress.[64]

Portuguese Factory and Settlement at Calicut

The new Zamorin who came to power in 1587 was friendly with the Portuguese.[65] The new viceroy, Dom Manoel de Sousa Coutinho, who succeed Dom Duarte de Menezes sent an armada against Kunhali.[66] The Zamorin was highly impressed with the naval might of the Portuguese. Padre Francisco da Costa, a Jesuit who was at that time a captive in the fortress of Kunhali succeeded in the liberation of himself and his companions from captivity and joined the other Portuguese.[67] In 1591 the Jesuits won the freedom to do missionary activities and the Zamorin agreed to a church in Calicut.[68] The Zamorin allowed the Portuguese to settle in Calicut. A Portuguese factor was allowed to look after trade endowed with the authority to issue passes to the ships. The factor who was looking after the affairs was under Belchior Ferreira.[69] The king himself laid the foundation of the Church in 1591 for which he generously granted not only the site but also the building materials. This was under a provisional treaty made with the Jesuit, Francisco da Costa.

During the time of Dom António de Noronha, Kunhali obtained permission from the Zamorin to establish a fortress near Pudupattanam, with a view to attacking ships loaded with pepper and passing through the Malabar coast. As per the terms of the treaty of 1591, the Zamorin was expected to keep pirates away from his land. Therefore he had the obligation of stopping the piratical activities of Kunhali too. Mahamet Kunhali Marakkar, the nephew of the previous Kunhali, succeeded him and started strengthening his settlement and fortress with more arms and ammunitions.[70] The Viceroy Mathias de Albuquerque sent Dom Alvaro de Abranches to the Zamorin cautioning him about the growth of Kunhali. Then an agreement with the Zamorin was concluded for joint action against Kunhali.[71]

After some time, the Zamorin violated the peace established with the Portuguese. The Portuguese confronted three ships of Calicut and killed more than two thousand people on board.[72] This was during the time of Viceroy Dom Mathias de Albuquerque. Later the Viceroy was told about the intention of the Zamorin to start war with Kunhali and to destroy his fortress.[73]The negotiations between the Portuguese and the Zamorin were not promptly concluded.

Andre Furtado, the captain in chief made an understanding in 1599 in the light of which the Zamorin promised to depute a number of chieftains, petty rulers like the princes of Tanor and Chaliam, and the administrator of Chaliam and so on to attack the fortress of Cunhale.[74] It was further agreed that as soon as the fortress of Cunhale was conquered, it would be immediately razed to the ground. The Captain promised to build a church in Calicut and to establish a factory there.[75] Again on 15 December 1605 the Portuguese made another agreement with the Zamorin and promised to have the church, and the priests at Calicut for spiritual service of the people there and also factor and other officials.[76]

Factory at Cochin

We find a totally different reaction to the Portuguese on the part of the king of Cochin. When Pedro Álvares Cabral faced great resistance in Calicut in 1500 on the part of the merchants there and the indifference of the king, he left for Cochin. It was reported by Gaspar da India that there was greater abundance of pepper than at Calicut though the king of the place was not as powerful as the Zamorin.[77] Though the Portuguese were afraid of the possible indignation of the king of Cochin on account of the capture of the ship belonging to the two merchants of Cochin (Mammale Marakkar and Cherina Marakkar), they were received warmly by the king in view of the setback suffered by the Portuguese at Calicut as informed by Miguel, a convert to Christianity at Calicut. Both Hindu and Muslim merchants of Cochin including Mammale Marakkar and Cherina Marakkar supported the king's intention to extend help to the Portuguese. Cabral asked Gonçalo Gil Barbosa to be the factor at Cochin, Lourenço Moreno and Bastião Álvares as writers, and Gonçalo Madeira from Tangir as interpreter and a few others to work in the factory. The Portuguese obtained large amounts of pepper and other spices from Cranganore also. Cabral was taken up by the generous gesture of the king of Cochin who promised to help them and fight for their interests even if his men would have to lose their lives. It was here that the Portuguese met a number of Christians of St.Thomas among whom a certain Mathias and Joseph. Both travelled in the fleet of Cabral to Lisbon, and from there to Rome, Jerusalem and Armenia to visit their Patriarch.[78]

The fleet under João da Nova which left Lisbon on 5 March 1501 had the information that there were two ports on the Malabar coast which would be friendly with Portugal. One was the port of

Cochin and so he made his way to Cochin.[79] João da Nova got this information through the letters left by Cabral at Mombassa on the African coast.[80] After paying a visit to the king of Cannanore he proceeded to Cochin on orders from the king. Portuguese officials like Gonçalo Gil Barbosa had been posted there by Cabral and so it was judged right to pick up cargos there.

Vasco da Gama led the fleet of twenty-five vessels to India in 1502. Five of them were for the protection of the Portuguese factories at Cochin and Cannanore. Diogo Fernandez Correa came as a factor for Cochin.[81] When Vasco da Gama came to Cochin in 1502, the representatives of the 30,000 Christians living in the vicinity of Cranganore visited him and surrendered a red rod of justice with three bells made of silver looking like a sceptre.[82] Vasco da Gama left Diogo Fernandez Correa as factor to stay safely in the wooden building made for this purpose. Thirty persons were also ordered to be with him there. Lourenço Moreno and Alvaro Vaz were appointed as writers to the factory. Finally Vasco da Gama left on 18 January 1503 for Cannanore and from there after signing the contract left for Lisbon where he arrived on 10 November 1503.

Afonso de Albuquerque and Francisco de Albuquerque with six vessels started from Lisbon on 6 and 14 April 1503 respectively. On arrival, Francisco de Albuquerque discussed with the king of Cochin the proposal of the Portuguese king for a fortress at Cochin. This received the approval of the king.[83] The king showed him a site for the construction of the fortress. The first Portuguese fortress was built in Cochin under the supervision of Francisco de Albuquerque, the brother of Afonso de Albuquerque. Afonso de Albuquerque who reached Cochin only by the end of September took up the work. The fortress made of timber was named Santiago and the church, after St. Bartholomeu.

The reception given by the king of Cochin to the Portuguese was not liked by the Zamorin. He prepared to oust the Portuguese from the Malabar coast and to fight against the king of Cochin. The armed men of the Zamorin took positions in Venduruthy to start the fight. Afonso de Albuquerque accompanied by 800 Portuguese soldiers and the prince of Cochin with 8,000 Nairs proceeded to face them. Duarte Pacheco was put in charge of the naval forces of the Portuguese. The Zamorin's forces comprised approximately

15,000 Nairs. The joint forces of the Portuguese and those of the king of Cochin valiantly fought and defeated those of the Zamorin. The principal Kaimals especially the five (later known as *anjikaimals*) who lived in the frontier areas of Cochin, paid their obeisance to the king of Cochin and promised to be loyal to him.[84]

The next calculated move came about in 1505 when Dom Francisco de Almeida, the first viceroy appointed, came to Cochin to establish the headquarters of Portuguese India. He was commissioned by the king to construct a strong fortress in Cochin and begin all governmental activities. He was asked by the king to annexe the territory of Cochin after the death of the then king, if possible.[85] When the Viceroy reached Cochin, the Tirumulpad (Trimumpara) or the ruler of Cochin being old had retired and the affairs of the kingdom were conducted by his nephew, *Nambeadiri* (*Nambeadora*).[86] Another of his nephews was fighting for the same position. He was not favourable to the Portuguese. Having been informed of the situation, Almeida decided to play his role as a diplomat and arranged a grand reception for the *Nambeadiri*. Setting aside the claims of the other one, he crowned him solemnly in Cochin with the golden crown the king of Portugal sent from Europe and sanctioned an annuity of 600 *crusados* per year to the king for the loss he had suffered in the fight with the Zamorin for the Portuguese.[87] The annuity was some sort of a compensation for the death of the two uncles of the king in the fight.[88] A written document saying that such an amount would be given to the king of Cochin in future was handed over to him. Then the viceroy requested permission for a strong masonry fortress in place of the wooden fortress. Though reluctantly, this was permitted. Being formally crowned by the Portuguese viceroy, the king of Cochin accepted the vassalage of the former. As suggested by the king of Cochin, the fortress was named Fort Manuel, or Manuel kotta in the local language. The formula of coronation and the investiture of the king of Cochin continued to be the same for several years.[89]

The friendly relations with the king of Cochin began to be shaken in the period between 1512 and 1516 on account of the treaty the Portuguese concluded with the Zamorin and the establishment of a factory and fortress in Calicut. The king of Cochin wrote letter after letter to the king of Portugal concerning the promises he had made to make him the greatest king of India and never to make any treaty of peace with the king of Calicut without his consent. [90] But there was no satisfactory response from him: So the king of Cochin tried to dissuade Affonso de Albuquerque from making any agreement with the Zamorin. Once the king knew that the treaty was ratified, his displeasure knew no bounds. In 1516 he wrote to the king of Portugal saying that it would be impossible for him to arrange spices at the rate fixed by the Portuguese. This was quite a blow to the market rate at Cochin. He alluded to the fact that spices were taken to places other than the Portuguese factories by the merchants on account of this.[91] The Portuguese went so far as to convert him to Christianity. However, he refused saying that this was a serious matter affecting the whole country.[92] Finally the Portuguese gave up the idea. Apart from this sort of occasional silent protest and renewal of friendship, nothing happened in the subsequent period that made the king withdraw his loyalty to the Portuguese.

Cannanore

When Pedro Álvares Cabral was loading his ships with pepper and other spices in Cochin in 1500, messengers from the kingdom of Cannanore met him and requested him to send the ships to the port of Cannanore to purchase spices and pepper. Cabral thanked the ruler of Cannanore and excused himself for not being able to take cargo from there at that time.[93] However, he informed the messengers that he would like to purchase some ginger from Cannanore on his way back to Portugal. The king of Cannanore sent two vessels in which one of the influential persons of the kingdom boarded and was asked to insist that Álvares visit Cannanore port. He offered to the Portuguese spices of all sorts at the port as well as peace and friendship. A number of vessels loaded with ginger and cinnamon appeared in Cannanore when Cabral reached there. He took only some ginger and cinnamon since his fleet was already loaded with required commodities from Cochin. He thanked the king for his generous offer of spices on credit and promised to take more cargo when the next fleet would be sent to India. The king of Cannanore, knowing that two people from Cochin were on their way to Portugal in the fleet of Cabral, sent his ambassador in that fleet to Portugal with some presents for the king of Portugal. Cabral left Cannanore on 16 January 1501 for Portugal.[94]

Even before the arrival of the fleet of Cabral in Lisbon, the king despatched another fleet of four ships on 5 March 1501 under João da Nova according to a royal decision to send a fleet every year to India. The fleet had the information that there were two friendly ports on the Malabar coast, Cannanore and Cochin. He made his way to Cannanore.[95] So João da Nova visited the king of Cannanore on his way to Cochin and was received cordially. He had orders from the Portuguese king that he should take cargo first from Cochin where the officials were posted by Cabral and on the way back to Lisbon to purchase commodities from Cannanore. João da Nova left five persons in Cannanore to start a factory there. Payo Rodrigues from the ship of Dom Alvaro, brother of the Duke of Bragança under Diogo Barbosa, the captain, was to be the head of the five persons left with the king of Cannanore. Another factor was from the side of Bartholomeu of Florence left by Captain Fernão Vinet.[96] He returned to Cannanore quickly after collecting the required cargo from Cochin. He collected commodities such as ginger, cinnamon and other spices from Cannanore and left a few more persons with Payo Rodrigues. He reached Lisbon on 11 September 1502.[97]

With a view to taking revenge upon the people of Calicut, Vasco da Gama set sail from Lisbon with a fleet of 25 vessels. Five of them were designed to be for the protection of the factories at Cannanore and Cochin. The fleet of Vasco da Gama brought the ambassadors of the king of Cannanore back home.[98] Vasco da Gama before proceeding to Cochin, visited the port of Cannanore and held discussions with king.[99] He wanted to get the prices of spices and pepper fixed at Cannanore. But he was annoyed with the words of two Hindu and two Muslim merchants sent by the local king in this regard since their attitude was totally different from that of the king. He left for Calicut and Cochin. Subsequently, on being informed of the indignation of Vasco da Gama, the king of Cannanore sent Payo Rodrigues to Cochin to inform Vasco da Gama about his willingness to reach an agreement regarding trade and commerce which would please the latter.

After having completed the assigned works in Cochin, Vasco da Gama came to Cannanore in January 1503. Vasco da Gama left

Gonçalo Gil Barbosa as factor at Cannanore besides Bastião Álvares and Diogo Godinho as writers and twenty men to take care of the needs of the factory. A contract with the local king was concluded.[100] The foundation stone for the fortress was laid by Francisco de Almeida in 1505. By 1507, when the Viceroy came to Cannanore from Angediva, the walls of the fortress were completed. Houses for eight men were built within the fortress. The church of St. James (Santiago) was constructed inside the fortress. There were several thatched houses (with *olas*) of earth near the fortress but close to the sea where the Portuguese guards resided.[101] An entrance from the bay to the fortress was also arranged for the people. The fortress was completed in March 1508. It was named 'Santãogil' according to the castle of St. Angelo in Rome.[102] A lot of pieces of artillery including those brought from Angediv were installed in the fortress. The Viceroy, convinced that the water at Cannanore was very good even for refining saltpetre and was useful for making ammunitions, asked Timoja of Honavar to send saltpetre to Cannanore.

Close to the sea, there was a hermitage under the invocation of Our Lady of Victory. As ordered by the viceroy a hospital was constructed next to it. Since Cannanore had a salubrious climate and water, for curing diseases contracted in the course of the voyage from Lisbon, the hospital attracted several sick people, especially those suffering from scurvy. The viceroy instructed the physicians at Cochin to send the sick to Cannanore.[103] By 1507 the ruler of Cannanore who established peace and friendship with the Portuguese was succeeded by his nephew. The new king was influenced by the Zamorin and turned against the Portuguese settled in the fortress. A ship belonging to one of the chief merchants of Cannanore loaded with horses from Ormuz was captured by the Portuguese though it had the *cartaz* or pass issued by them.[104] This gave rise to unrest in Cannanore. The fortress was besieged by the people and there ensued a furious fight between the local people and the Portuguese. With the assistance of the Zamorin the king of Cannanore put up a stiff fight with the Portuguese. After having been under siege, the Portuguese succeeded in defeating the king who became close to them for the rest of his reign. A treaty of peace[105] between the king of Cannanore and the Portuguese was signed with the help of Tristão da Cunha.[106] With the exception of the trouble in 1507 the relations were fairly cordial and the fortress and the factory at Cannanore functioned satisfactorily. The king wrote to Portugal in 1516 to see that horses from Gujarat and ropes from Ormuz came to Cannanore in abundance.[107]

Quilon

When Pedro Álvares Cabral was loading his ships with pepper and other spices in Cochin in 1500, messengers from the kingdom of Quilon offering spices and pepper came to Cabral. He thanked the ruler of Quilon and excused him for not being able to take cargo from there that time.[108] But when Vasco da Gama came to Cochin after taking revenge on Calicut in 1502, the Queen of Quilon again requested him to sends two ships to Quilon to buy spices. Da Gama willingly sent the vessels. A certain Mathias, a Christian from Kayamkulam helped the Portuguese load the ships with spices.[109] Again in 1503 when Afonso de Albuquerque and Francisco de Albuquerque were busy loading cargos in Cochin, the Queen of Quilon sent messengers requesting them to send two ships to Quilon to collect commodities from there. Afonso de Albuquerque went to Quilon, concluded a treaty with the ruler and collected required cargo from there.[110] The ruler gave him a grand reception with all the solemnities. During the interview, the king promised to give Albuquerque all the spices he needed and signed a treaty of friendship with the Portuguese. Albuquerque established a factory there in 1503 itself.[111] Though with reluctance the king gave in to the request for accepting a Portuguese man to look after the matters of administration of justice for the Portuguese left behind in Quilon as well as for the local Christians who, according to the report of the eyewitness amounted to 3,000.[112] Antonio de Sá de Santarém was left there as factor along with two writers, Ruy de Araujo and Lopo Rabello besides twenty men to look after the needs of the factory.[113] The agreement was inscribed on a silver plate. Mathias from Kayamkulam helped the Portuguese to obtain their required cargos. He and his brother always remained committed to the help of the Portuguese. They were jubilant about the arrival of the Portuguese at Quilon especially on account of the shared Christian faith.[114]

Quilon was the principal town of Malabar before the rise of

Calicut. It was the richest town of the entire Malabar coast. It had trade relations with Coromandel, Ceylon, the Maldives, Bengal, Pegu, Sumatra and Malacca. It had a very miraculous church built by St.Thomas (known as *Martama, Martoma*) the apostle when he went there to preach the Christian faith.[115]

Dom Francisco de Almeida while in Cochin in November 1505 received a message from Quilon regarding the murder of the Portuguese factor António de Sá, the other Portuguese men and the destruction of the church of St. Thomas in Quilon. He sent his son D. Lourenço to chastize the people involved in the case.[116] After having put the Muslims of Quilon to flight, he returned to Cochin. In fact Dom Francisco de Almeida was against setting up a fortress at Quilon. He was of the opinion that larger the number of fortresses in India, the weaker would be the Portuguese power in India. For them the entire force had to be concentrated on the sea in view of fight with the Venetians and Turkish sultan.[117]

Damages caused to the Portuguese were repaired in the light of a treaty of peace and commerce signed by the ruler of Quilon and Governor Lopo Soares in the name of the king of Portugal on 25 September 1516.[118] Irnacalao (Iravimangalam?), the king of Desinganat (ancient name of the kingdom of Quilon), Caycoy Irnacalao, his sister and the governors agreed to rebuild the church of St.Thomas destroyed in 1505, to deliver 500 *bhars* pepper as compensation for the death of the Portuguese factor, and to give all the spices in future to the Portuguese factory at the same price as in Cochin and not to levy any customs duties on the trade conducted by the Portuguese. According to the terms, if anybody wanted to conduct trade, he had to take a permit from the captain of Cochin or the captain-in-chief. In the matter of administration of justice, the Portuguese were given some sort of an exemption in the sense that if the natives of Quilon were found guilty on account of their quarrels with the Portuguese or the local Christians, judgement would be pronounced and punishment would be executed by the Portuguese captain. In the light of the secret agreement concluded with the queen of Quilon, Eytor Rodrigues started the construction of a factory and a fortress in September 1519.[119] The fortress was named St.Thomas. The treaty with the queen signed in 1519 was in favour of the Portuguese.[120]

In the same year about 5,000 bullock-loads of pepper taken from Quilon along Ariankavu pass were captured by the factor since the ruler refused to intervene saying that it was the wealth of Brahmins (*Brahmaswam*).[121] The people of the locality besieged the fortress and in the subsequent encounter the Portuguese won. Another treaty was concluded on 17 November 1520.[122] The queen was obliged to deliver the pepper due to the Portuguese immediately and all the Portuguese having trade in Quilon were expected to give customs duties to the ruler of Quilon as in Cochin. The captain of the fortress was to issue *cartazes* to the merchants of the locality. No pepper was to be sold to anyone except the Portuguese factor.[123] This treaty continued be in force till 1543. On 25 October 1543 Governor Martim Afonso de Sousa signed another contract with the ruler of Quilon containing the details of trade and the terms regarding the treatment of Portuguese and native criminals.[124]

Fortress at Cranganore

Cranganore was the entrance for the Zamorin and his people to the principality of Edappally where the coronation ceremony used to be conducted. Edappally was attached to the kingdom of Cochin. On account of the strategic importance of Cranganore, Dom Francisco de Almeida back had in 1508 written to the Portuguese king that it would be good to have a fortress at Cranganore by the side of a river which had connection to Calicut. If this was done, the flow of pepper through this river could be completely stopped.[125] This was necessitated by prevailing political conditions. As indicated by Diogo de Couto the entire population of Malabar was divided into two camps, the one that favoured the Zamorin and were generally called *paydaricuros* and the other that supported the king of Cochin and were called *logiricuors*.[126] The powerful people on the side of the king of Cochin, besides a large number of Christians who were the descendants of those converted by St.Thomas living in the vicinity would be the vassals of the king of Portugal.

It was in 1536 that Martim Afonso de Sousa issued orders to set up a fortress in Cranganore. Diogo Pereira was given charge as the captain. He started work immediately and had been given gunners and twenty men. Subsequently a college for the training of the local clergy was also set up. The college was founded by Frei Vicente de Lagos, a member of the Franciscan province of *Piedade* in 1540. This was meant exclusively for the training of the clergy for the St.Thomas Christians.[127] According to the report of Simão Botelho, by the middle of the sixteenth century there were a captain, a writer for the factory, a parish priest, a local cleric, a college for the training of the clergy, and a church under the invocation of St. James at Cranganore.[128] The castle was called St.Thomas. In 1574 this fortress was attacked and conquered by the Dutch in 1662.

Barcelor

The Portuguese Viceroy Dom Luis de Taide conquered the fort of Barcelor in 1568 after conquering Honavar or Onor. This fortress was then taken by Sivapa Naique, the king of Bednur in August 1652.

Mangalore

Viceory D. Antão de Noronha constructed the fort of Mangalore in 1568 and named it St. Sebastian because the foundation stone for it was laid on 20 January 1568. Sivappa Naik, the king of Ikkeri took over the fortress in September 1653.

Goa

Afonso de Albuquerque seized Goa from Adilshah of Bijapur and converted it into a Portuguese territory in 1510. He ordered a church under the patronage of St. Catherine to be built in Goa next to the hospital. Diogo Fernandes, the factor was asked to look after the construction of the church. The headquarters of Portuguese India set up in Cochin in 1505 was shifted to Goa in 1530. The Church of St. Catherine was reconstructed in 1532. When João de Albuquerque came as Bishop nominated to the See of Goa in 1539, it was elevated to the status of a cathedral. Until 1542 the cathedral was the only parish church in the city. Pope Paul VI elevated the cathedral to the Archiepiscopal and metropolitan See. Viceroy D. Francisco Coutinho ordered on 4 November 1562 the construction of a bigger cathedral with the money raised through the sale of properties confiscated from Hindus and other non-Christians who died without a testament or heirs. A site slightly away from the church of St. Catherine was chosen for this. The main section of the church was completed in 1619 and the rest of the buildings in 1631. The architects were Ambrosio Argueiros and Julio Simão.

Fortress of Bardes

The fortress of Bardes named *Reis Magos* was constructed on the ruin of a castle of Adil Khan by Viceroy Afonso de Noronha (1551-4). Improvements were made by the governors Manuel de Souza Coutinho and Caetano de Melo Castro

Fortress of Rachol

The fortress of Rachol was constructed in 1535 in Salcete.

Chaul

Chaul is situated on the north or right bank of the Kundalika River where it meets the sea. It was under the nominal control of the sultans of Ahmednagar. Burhan Nizamshah of Ahmednagar (1508-53) permitted the Portuguese to build factory at Chaul in 1516. Chaul was under the Nima-ul-Mulk of Ahmednagar. Governor Diogo Lopes de Sequeira got the fortress constructed here in 1521. A Franciscan monastery dedicated to Sta. Barbara was established in Chaul by Frei Antonio do Porto who visited Chaul in 1534. There were two settlements, one known as upper Chaul and the other as lower Chaul. The latter was wholly dominated by the Portuguese. Having secured the northern bank of the river by building the Revdanda fort, the Portuguese tried to get hold of the ridge of Korlai in which they succeeded only after a bitter battle.

The Portuguese fortress of Revdanda was located in lower Chaul, whose circumference is about 5 km. An inscription records that the factory was established in 1516 and that the fortifications around it were made in the period 1521-4.

The village of Korlai at the base of a rocky, 100-m high ridge, was fortified and was known as the Korlai fort. The Portuguese called it the Cheul Rock-Morro de Chaul. They had realized the strategic importance of the place and wanted to build their own fort there. With a fort on either side of the Kundalika River, they would have been able control all river and sea traffic in that area. Their request was, however, refused by Hussain Nizamshah. To pre-empt any mischief on the part of the Portuguese, Burham Nizamshah built a small fort and placed Fateh Khan in command. He tried to bombard Revdanda on the opposite shore, but was not successful. On the contrary the Portuguese attacked and occupied the fort and renovated it around 1594. It successfully repulsed the attack by the Marathas under Sambhaji in 1683. The fort was about 100 m long but only 30 metres wide at its widest point. There are seven terraces, each protected by a wall and a gate. Ruins of a church can be seen at the top. The fortification at the north-eastern end reached down to the sea. At one time this was an important Portuguese fort manned by 8,000 men and protected by 70 guns. An inscription above a doorway on the highest part of the fort records the building of the fort in 1646. Subhanji Mankar who was specifically sent with the mission by Chimaji Appa in 1739-40 took over the fort. Portuguese Chaul was ceded to the Marathas in view of the treaty of Poona dated 18 September 1740.

Dabhol

It was occupied by the Portguuese in 1508. The Portuguese writers speak of a well-fortified fort at Dabhol. It was a centre of maritime trade.

Bombay

The Portuguese had a manor house on the main island and a small fort at Mahim. The Portuguese did not pay great attention to Bombay since it was almost uninhabited. It was given as a part of the dowry of Catherine of Bragança when she was married to King Charles II of England in 1661. By 1665 the English East India Company took possession of the Bombay Island from the Portuguese.

Tarapur

Tarapur is located on the south bank of the Tarapur creek which was situated between Bassein and Daman. In view of the treaty with the sultan of Gujarat, the Portuguese were permitted to build a massive fortress here. It was attacked by the Mughals and the Siddis. But it was not conquered until it fell to the Marathas in 1738-9.

Mahim

Mahim has been always linked with Kelve (a small village) and was called Kelve Mahim to distinguish it from Mahim Island. It was taken over by the Portuguese who built a fort on an earlier structure. In 1739 it passed into the hands of the Marathas. Ruins of a Portuguese church are visible in the Kelve fort.

Agashi

Agashi was a famous centre of shipbuilding. Some historians believe that there was a Portuguese fortress in Agashi, though no traces are now visible. The Portuguese built a church there in 1540. The community was very large.

Arnala

It is an island south of Agashi Bay. It was known as *Ilha de Vacas* during the period of the Portuguese sway. The fort was built in 1516 by Muhammed Begeda of Gujarat and later in 1536 it passed into the hands of the Portuguese . It was an important naval and military station under the Portuguese. The Marathas occupied it in 1739.

Dahanu

It is situated 6 km from the Dahanu Railway Station on the Western Railway. The port is on the northern bank of the Dahanu creek. There are Motha (big) and Chota (small) Dahanu. The fort is located in Motha Dahanu. The fort was constructed by the Portuguese in 1533-4 on the northern bank of the creek.[129] The fort had a church in the name of *Nossa Senhora das Angustias*

Basssein

Bahadur Shah the sultan of Gujarat ceded Bassein to the Portuguese through a treaty of 23 December 1534 on board the ship *S. Mateus* which had appeared at the port of Bassein. Francisco Barreto the Portuguese governor annexed the fortresses of Asserim and Manora. Bassein functioned as the capital of the Northern Province of Portuguese India. There were a cathedral, five convents, 13 churches, one orphanage and a hospital in and around the fortress under the Portuguese. It passed on to the Marathas in 1739 after a heroic battle waged by the Portuguese against the Maratha forces.

Diu

The sultan of Gujarat ceded Diu to the Portuguese in view of the treaty of 25 October 1535. On 20 November 1535 the governor laid the foundation stone of the fortress in Diu which was christened as the Fortress of St. Thomas. Twice the fortress was sieged, once in 1538 and then in 1546. The customs house in Diu yielded a large sum a part of which was sent to Goa.[130]

Daman

The fortress of Daman was ceded to the Portuguese by the sultan of Gujarat. Viceroy D. Constantino Bragança took possession of it on 2 February 1559. The Portuguese set a fortress there which was named after Our Lay of Purification. It was designed by the nephew of the Archbishop of Braga, Dom Frei Bartholomeu dos Martires who learned architecture in Flanders.

The Coromandel Coast

The Portuguese contacts with Mylapore developed after the discovery of the tomb of St. Thomas in 1517. Many Portguuese had already settled down at Mylapore even before the church of St. Thomas was rebuilt in 1524.The Portuguese on account of the flourishing commerce prevailing in Pulicat set up the earliest Portuguese trading station on the Coromandel coast at Pulicat. Manuel de Frias was appointed Captain of Coromandel in 1522 and he resided in Pulicat. It had an office to issue passes or *cartazes*. Later the seat of the Captain of Coromandel was shifted from Pulicat to Santhome at Mylapore. The next captain, Miguel Pereira, took up his residence there in 1530.

The Portuguese had settled in places like Pulicat, Santhome, Devanampattinam, Porto Novo, Tranquebar (Tarangambadi) and Nagapattinam. A number of Portuguese people settled in these places with their families. They did not bother about the government. They conducted their own private trade in association with local merchants. They were averse to the royal control and the appointment of a captain by the Portuguese king for the Coromandel coast.

The Portuguese enclave in Nagapattinam came into existence in 1594. The Portuguese exported rice from here to Ceylon. The Naik of Thanjavoor kept good relations with the Portuguese at Nagapattinam. This settlement did not have any fortress since Naik did not give his consent for in. The Portuguese at Nagapattinam had commercial relations with Malacca, Manila, and also with the Portuguese factories on the western coast. They used to have commercial contact with Orissa, Bengal, Pegu, and so on. The Dominican and Franciscan friars, Augustinians and the Jesuits had houses at Nagapattinam in the seventeenth century. There was a church of Mercy (*Igreja da Misericordia*).

Santhome of Mylapore

The settlement was surrounded by a wall on the side of the sea (Bay of Bengal) with houses having their entrance from the seashore. The town was sanctified by the house of St. Thomas, the Apostle. The Portuguese settlement was established here on account of the church of St.Thomas. It had three bulwarks, of St. Dominic, St. Paul and St.Thiago (James) on the seaside. The other bulwarks were those of Antonio da Costa, St. Augustine, bulwark of ironsmiths, bulwarks of Salvador Rezende, João Roiz de Souza, and Mother of God. There were thirty pieces of artillery on all the bulwarks. There were many Christians amounting approximately to 6,000 outside the city around 1612. The convents and churches of St. Dominic, St. Augustine and St. Paul under the Jesuits, the church of our Lady, the Church of St. Francis under Franciscans, Our Lady of Luz, Mother of God and of St. Lazar were at the service of the Christians in Mylapore. The cathedral church, the See, was that of the glorious Apostle St. Thomas constructed on the pagoda where there was a small chapel made of wood. Being very heavy the latter could not be lifted from the water by a number of persons, but which St.Thomas managed to take it out from the water to the land by the chord tied around his waist. There was a church of Our Lady on the big Mountain where St. Thomas suffered his martyrdom. There on a marble stone was found a cross sculptured from the same rock by the hand of the apostle. There was yet another church of similar construction on the small mountain where he used to be present. There was a house of devotion and a hole in a rock through which the Saint went out when they attempted to kill him. There are crosses sculptured all over the rock.

Bengal

João de Silveira was the first Portuguese commander of an expedition to Bengal in 1517. He landed on the coast of Arakan and then steered to Chittagong which was the chief port and the main gateway to the royal capital Gaur. It was situated at the mouth of the river Meghna and was convenient for navigation. With the fall of Gaur, Chittagong too lost its importance and trade was diverted to Satgaon, which in its turn was supplanted by Hoogly. All the Portuguese who came to Bengal first entered Chittagong. They named it *Porto Grande* in contrast to Satgaon called as *Porto Pequeno.* Satgaon situated on the river Saraswati, was the chief port of Bengal. Large vessels could come to Satgaon. The towns of Hoogly, Chandernagore, Chinsura and Serampore did not even exist in name. Afonso de Albuquerque informed the Portuguese about the possibility establishing commercial relations with Bengal. The king commissioned in 1517 Fernão Peres de Andrade with four ships particularly to open a trade with Bengal and China. But he could not reach Chittagong as planned. But João Coelho sent by Fernão Pires arrived at Chittagong before Silveira came in an expedition. Silveira sent a messenger to the king of Bengal asking permission to start a factory and trade with Bengal. The messenger was never received. João Coelho was well received by the governor of the ruler of Bengal, but Silveira was kept away. So, Silveira went to Arakan which was subject to the king of Bengal. There too he could not succeed in establishing a treaty of peace and commerce.

When Sher Shah planned out the invasion of Bengal through

the passes of Teliagarhi and Sikligali, the gateways to Bengal, the Portuguese offered a stubborn resistance in 1536 and prevented Sher Shah from taking the city of Ferranduz, about 20 km from the city of Gaur. Mahmud Shah of Bengal was very pleased with the help. On account of their help Mahmud Shah permitted the Portuguese to set up factories in Chittagong and Satgaon in Hoogly district and to collect taxes. These were the earliest Portuguese establishments in Bengal. Again in 1538 Sher Shah organized another campaign against Bengal and invaded Gaur. Humayun advanced against Sher Shah and forced the latter to retreat from Gaur to Sasseram.

Hoogly was not at all significant as Satgaon during this period. The first settlement of the Portuguese was in Satgaon, not in Hoogly. The second settlement was founded in Hoogly proper by Pedro Tavares to whom Akbar granted a *farman* (1579-80). The third settlement was in Bandel, close to the previous one, under the *farman* of Shah Jahan granted in 1633. The Augustinians had by 1599 built a great convent in Bandel. Towards the latter part of the sixteenth century, the greater part of Bengal trade had passed into the hands of the Portuguese. Besides Hoogly, Satgaon, and Chittagong, they had also trade in Hijili, Banja, Dacca and many other small ports. When the Portuguese settled in Hoogly, they diverted all the trade to their own port to the detriment of Satgaon and gradually developed Hoogly.

Thus we can say that the Portuguese were established by the end of the sixteenth century in very important and strategic points on the west and eastern coasts of India. But we do not come across as many well-fortified settlements on the eastern coast as we find on the western coast. We further notice that installations for ship-building were chiefly set up on the western coast

NOTES AND REFERENCES

1. C.R. Boxer, *The Portuguese Seaborne Empire 1415-1825*, London, 1973, p. 1
2. João de Barros, *Da Asia,* Decada I, part I, Lisboa, 1778, p. 184
3. *Ibid.*, p. 194.
4. *Ibid.*, pp. 194-6.
5. *Ibid.*, pp. 267-8
6. For detailed studies of Vasco da Gama, see A.C. Teixeira de Aragão, *Vasco da Gama e a Vidigueira: Estudo histórico*, Lisboa, 1898; K.G. Jayne, *Vasco da Gama and His Successors 1460-1580,* London, 1910; Henry E.J. Stanley,

The Three Voyages of Vasco da Gama and His Viceoryalty, London, 1869; Armando Cortesão, *The Mystery of Vasco da Gama*, Coimbra, 1973; Sanjay Subrahmanyam, *The Career and Legend of Vasco da Gama*, Cambridge, 1997. Barros, *op. cit.*, Decada I, part I, p. 270.

7. *Memoria das armadas que de Portugal passaram ha India e esta primeira e ha com que Vasco da Gama partio ao descobrimento de la por mandado de El Rey Dom Manuel no segumdo anno de seu reinado e no do nacimento de xto de 1497* (Lisboa, Facsimile edition, p. 1; Barros, *ibid.*, Decada I, part I, p. 279.
8. Barros, *ibid.*, Decada I, part I, p. 328.
9. João Pedro Garcia et al., *Vasco da Gama e a Índia*, Lisboa, 1998, p. 25.
10. W.W. Rockhill in his *Notes on the Relations and Trade of China in T'ong-Pao,* vol. XV (Leiden, 1914), p. 425, alludes to Chinese trade with this coast in 1296 and mentions Panam and Fandaraina among the ports alluded to in the *Yuan Shih.* Fandaraina or Pantalayani seems to be mentioned also in another Chinese authority of the same period (*ibid.*, p. 435, note 1).
11. Mahdi Hussain, ed. & transl., *Rehla of Ibn Batuta*, Baroda, 1976, p. 188.
12. *Ibid.*, p. 188.
13. 'When we arrived at Calicut the king was fifteen leagues away (in Ponnani). The captain major sent two men to him with a message informing him that an ambassador had arrived from the king of Portugal with letters and that if he desired it he would take them to where the king then was. … He sent word to the captain bidding him welcome, saying that he was about to proceed to Qualecut. As a matter of fact, he started at once with a large retinue' E.G. Ravenstein, ed., *A Journal of the First Voyage of Vasco da Gama 1497-99,* Delhi, 1995, p. 50.
14. 'A pilot accompanied our two men, with orders to take us to a place called Pandarani, below the place (Capuano) where we anchored at first. At this time we were actually in front of the city of Calecut. We were told that the anchorage at the place to which we were to go was good, whilst at the place we were then it was bad, with a stony bottom, which was quite true, and more over that it was customary for the ships which came to this country to anchor there for the sake of safety. We ourselves did not feel comfortable, and the captain-major had no sooner received this royal message than he ordered the sails to be set, and we departed. We did not, however, anchor as near the shore as the king's pilot desired'. Ravenstein, ed., *A Journal of the First Voyage of Vasco da Gama 1497-99,* p. 50. The Portuguese fleet remained at Pandarane from 31 May to 23 June 1498. On 24 June, they took the merchandise to Calicut.
15. 'Passing thereby is another town on the coast called Tircore and passing this there is another which they call Pandarane beyond which there is yet another with a small river which they call Capucate. This is a place of great trade and many ships, where on the strand are found many soft

sapphires' Duarte Barbosa, *The Book of Duarte Barbosa*, vol. II, London, 1921, pp. 85-6.

16. Tomé Pires, *The Suma Oriental of Tomé Pires*, Delhi, 1990, vol. 1, pp. 74, 78.
17. Barros, *op. cit.*, Decada I, part I, p. 335.
18. *Ibid.*, p. 357.
19. *Ibid.*, p. 370.
20. *Ibid.*, p. 371.
21. *Ibid.*, pp. 372-7.
22. *Ibid.*, pp. 378ff.
23. *Ibid.*, p. 384.
24. *Cronica do Descobrimento e conquista da India pelos Portugueses,* Coimbra, 1974, p. 21.
25. Fracanzano Montalbodo, *Paesi Nouvamente Retrovati & Novo Mondo da Alberico Vesputio Intitulato*, Venezia, 1597, Facsimile, London, 1916, p. 90.
26. Montalbodo, *op. cit.*, p. 96; Constancio Roque da Costa, *Historia das Relações diplomaticas de Portugal no Oriente*, Lisboa, 1895, p. 22, Leonardo da Ca' Masser, '*Relazione . . .*', in *Archivio Storico Italiano*, Appendice, tom II; Firenze, 1845, pp. 15-16.
27. Thomé Lopes, 'Navegação as Indias Orientaes', in *Colecção de Noticias para a Historia e Geografia das Nações ultramarinas que vivem nos Dominios Portugueses ou lhes são visinhos*, tom II, nos. 1 & 2, Lisboa, 1812, p. 190, 'Eu vim a este porto com boa mercadoria, para vender, comprar e pagar os vossos generos; estes são os generos desta terra, eu vo-los envio do presente, comoe também as el Rei', also Barros, *Da Asia,* Decada I, part II, p. 53.
28. Fernão Lopes de Castanheda, *Historia do Descobrimento e Conquista da Índia pelos Portugueses*, Livro I, Coimbra, 1924, p. 233.
29. *Cartas de Albuquerque*, Tomo, Lisboa, 1884, p. 250.
30. *Cartas* I, p. 152, Affonso de Albuquerque writes that it was the brother of the deceased Zamorin called Nambiadiry who expressed his willingness to come to terms with the Portuguese. Castanheda too writes that Nambeadiri was his brother. Fernão Lopes de Castanheda, *Historia do Descobrimento & Conquista da Índia pelos Portugueses,* Livro III, Coimbra, 1928, p. 291.
31. *Cartas*, I, p. 250.
32. *Ibid.*, p. 137.
33. Gaspar Correa, *Lendas da India*, tomo II, Coimbra, 1924, p. 330, *Cartas*, I, p. 152.
34. *Cartas*, I, p. 152.
35. *Ibid.*, p. 153.
36. Julio Firmino Judice Biker, *Collecção de Tratados e Concertos de Pazes que o Estado da India Portuguez fez com os Reis e Senhores com quem teve Relações*

nas partes da Asia e Africa Oriental desde o Principio da conquista até ao fim do Seculo XVIII, Lisboa, 1881, pp. 22-3.

37. Biker, *Tratados*, pp. 28-33.
38. Correa, *Lendas*, tomo II, p. 334.
39. *Cartas*, vol. VII, p. 131.
40. Hermann Gundert, ed., *Kerala Palama (1498-1531)*, Kottayam, 1959, p. 145.
41. *Kerala Palama*, p. 166.
42. The *Tohufut-ul-Mujahideen*, p. 117; Gaspar Correa gives details about the fight put up by the Zamorin of Calicut against the Portuguese and the fortress at Calicut, ref. Correa, *op. cit.*, tomo II, part II, pp. 810ff., 890-918; *ibid.*, pp. 941-64.
43. R.S. Whiteway, *The Rise of the Portuguese Power in India 1498-1550*, London, 1899, p. 204.
44. Barros, *Da Asia*, Decada IV, part II, Lisboa, 1973, rpt., pp. 451-2, Duarte Barbosa describes the fortress. 'Two leagues beyond this place (Capucate) is the city of Calicut where in more trade was carried on, and yet is, by foreigners than by the natives of the land, where also the king our Lord, with the full assent of the king thereof, holds a very strong fortress. To the south of this city there is a river on which lies another town called Chiliate, where dwell many moor, natives of the land who are merchants and have many ships in which they sail'. . ., Duarte Barbosa, *The Book of Duarte Barbosa*, London, 1921, vol. II, pp. 86-7. The fort was built on the right bank of the Kallayi River at the southern end of the town close to the old jetty stormed by Albuquerque in 1510. In shape and size it was exactly like the Cochin fort. On the sea side there were two towers and the wall connecting them was pierced by a wicket gate so that the garrison might have easy and uninterrupted communication with the sea. The keep had three storeys. On the land side also there were towers and between them was the principal entrance of the fort defended by bastion. Barbosa, *op. cit.*, p. 87 footnote.
45. Fernão Lopes de Castanheda, *História do Descobrimento e Conquista da India pelos Portugueses*, Coimbra, 1924, livro VIII, p. 270. An amount of 1,000 golden *pardaos* was given to the ruler for the consent.
46. Barros, *Da Asia*, Decada IV, part I, pp. 470-5; Simão Botelho, 'O Tombo de Estado da India, in Rodrigo José de Lima Felner, *Subsidios para Historia da India Portugueza*, Lisboa, 1868, pp. 130-2. Diogo de Couto, *Da Asia*, Decada IV, Lisboa, 1973, rpt., part II, pp. 196 ff. Chaliyam is an island formed by the Beypore and Kadalundi rivers, held by the Portuguese after they left Calicut in 1525. 'A mound where stood the Portuguese fort destroyed by the Zamorin in 1571 is still visible at the sea's edge.' Barbosa, *op. cit.*, vol. II, p. 87, footnote.This fortress at Chale/Chaliyam was called Santa Maria do Castello, ref. Correa, *op. cit.*, tomo III, part I, pp. 435-7.

47. Couto, *op. cit.*, Decada VI, part II, pp. 210-11 gives the nature of the work of *Jangada* or *Changathi* .
48. Correa, *op. cit.*, tomo III, part I, p. 435 .The details of the structure are furnished by Correa, *ibid.*, pp. 437-8. He says that there was no problem for this fortress till 1563 when he was writing the history. The plan of the fortress is given by Correa in Tomo II, part II, before page 439. The plan shows a church and houses outside the fortress.
49. Castanheda, *op. cit.*, livro VIII, pp. 429-36.
50. *Ibid.*, livro IX, p. 560. Castanheda gives the details of the agreement.
51. Biker, *Tratados,* pp. 88-94; Simão Botelho, *O Tombo de Estado da India*, pp. 249-54.
52. Couto, *Da Asia,* Decada VII, part II, p. 495.
53. *Ibid.*, pp. 516-18.
54. Couto, *Da Asia,* Decada VIII, p. 459.
55. *Ibid.*, Decada IX, p. 9.
56. For details of the route of trade in spices before the arrival of the Portuguese, and the advantage reaped by the Sultan of Cairo through trade in this route. Fernão Lopes de Castanheda, *Historia do Descobrimento & Conquista da India pelos Portugueses*, Coimbra,1924, livro. 1, pp. 381-3.
57. Castanheda, *História do Descobrimento & Conquista da India pelos Portugueses*, livro. 1, pp. 356-9.
58. Couto, *Da Asia,* Decada X, part II, p. 144.
59. *Ibid.*, pp. 27-9.
60. *Ibid.*, pp. 148-9.
61. *Ibid.*, pp. 186-93.
62. *Ibid.*, pp. 190-3.
63. *Ibid.*, p. 315.
64. *Ibid.*, pp. 340-3
65. Couto, *Da Asia,* Decada VII, part II, p. 528. He writes that this Zamorin continued to rule till 1610.
66. *Ibid.*, Decada XI, pp. 72 ff.
67. *Ibid.*, 72-3.
68. *Ibid.*, pp. 73, 184.
69. *Ibid.*, Decada XII, p. 70.
70. *Ibid.*, Decada XI, pp. 185-6.
71. *Ibid.*, p. 188.
72. *Ibid.*, p. 177.
73. *Ibid.*, Decada XII, p. 67.
74. For details of the fortress and the way in which Cunhali was caught and beheaded see Pyrard de Laval, *The Voyage of François Pyrard of Laval*, vol. 1, London, 1887, pp. 350-8; Couto, *Da Asia*, Decada XII.

    The plan of the fortress of Kunhali at Kottackal is reproduced in Francisco Pyrard de Laval, *The Voyage of François Pyrard of Laval*, vol. 2, part II, between pp. 510 and 511.This is reproduced from *Livro da India* of P. Barreto de Resende, Sloane Collection, no. 197.

75. Biker, *Tratados*, pp. 186-8; Couto, *Da Asia*, Decada XII, liv. IV, capt. II.
76. Biker, *Tratados*, p. 189.
77. Barros, Decada I, part I, pp. 440-1.
78. *Ibid.*, pp. 446-7. Mathias died after visiting Lisbon while Joseph proceeded to Rome and Venice. He described at Venice the customs and manners of the Christians of St.Thomas Christians, a summary of which has been incorporated in the Latin Book entitled *Novus Orbis*.
79. *Ibid.*, p. 467.
80. *Ibid.*, p. 466.
81. Barros, Decada I, part II, p. 23.
82. *Ibid.*, p. 63.
83. Correa, *Lendas da India*, pp. 384-93.
84. *Ibid.*, pp. 387-93.
85. Raymundo António de Bulhão Pato, ed., *Cartas de Affonso de Albuquerque*, tomo II, Lisboa, 1898, pp. 323.
86. João de Barros, *Da Asia*, Decada I, part II, Lisboa, 1777, p. 351.
87. Castanheda, *História do Descobrimento & Conquista da Ìndia pelos Portugueses*, Livro I, Coimbra, 1924, p. 255; João de Barros, *Da Asia*, Decada I, part II, Lisboa, 1777, pp. 355-6.
88. Raymundo António de Bulhão Pato, ed., *Cartas de Affonso de Albuquerque*, tomo III, Lisboa, 1903, pp. 73-84; Barros, *Da Asia*, Decada I, part II, Lisboa, 1777, p. 356.
89. Silva Rego, António da, ed., *Documentação Ultramarina Portgesa*, III, Lisboa, 1963, pp. 355 ff.
90. Raymundo António de Bulhão Pato, ed., *Cartas de Affonso de Albuquerque*, tomo III, Lisboa, 1903, pp. 39, 73-84; Barros, *Da Asia*, Decada I, part II, p. 356.
91. Raymundo António de Bulhão Pato, ed., *Cartas de Affonso de Albuquerque*, tomo IV, Lisboa, 1910, pp. 71-3.
92. Raymundo António de Bulhão Pato, ed. *Cartas de Affonso de Albuquerque*, tomo I, Lisboa, 1884, pp. 367-9.
93. Barros, Decada I, part I, p. 448.
94. *Ibid.*, pp. 456-8.
95. *Ibid.*, pp. 467.
96. *Ibid.*, p. 473.
97. *Ibid.*, p. 478.
98. Barros, Decada I, part II, p. 24.
99. *Ibid.*, p. 39-40.
100. *Ibid.*, pp. 74-5; *O Tombo do Estado da India* por Simão Botelho, pp. 28-9.
101. Castanheda, *Historia do Descobrimento & Conqista da India pelos Portugueses*, livro I, p. 307

102. Gaspar Correa, *Lendas da India*, tomo I, part I, Coimbra, 1922, pp. 583, 728.
103. Correa, *Lendas,* pp. 729-30.
104. Raymundo António de Bulhão Pato, ed., *Cartas de Affonso de Albuquerque*, Lisboa, 1898, tomo II, p. 401
105. Barros, Decada II, part I, pp. 62-76, Damião de Gois, *Cronica do Felicissimo Rei D. Manuel,* part II, Coimbra, 1953, pp. 50ff., *Cronica do Descobrimento e Conquista da India pelos Portugueses*, Coimbra, 1974, pp. 158-9.
106. Castanheda, *História do Descobrimento*, pp. 304-23.
107. João de Souza, *Vestigios da Lingua Arabica em Portugal, ou Lexicon Etymologico des Palavras e nomes Portugueses que tem origem arabica*, Lisboa, 1789, p. 81.
108. Barros, Decada I, part I, p. 448.
109. Raymundo António de Bulhão Pato, ed., *Cartas de Affonso de Albuquerque*, tomo II, Lisboa, 1898, p. 268.
110. Castanheda, *História do Descobrimento*, p. 127.
111. Correa, *Lendas da India*, tomo I, part I, p. 407.
112. Giovanni da Empoli, 'Viagem as Indias Orientaes', in *Collecção de Noticias para a Historia e Geografia das nações ultramarinas, que vivem nos dominios portugueses ou lhes são visnhos*, tomo II, nos. 1 & 2, Lisboa, 1812, pp. 224-7.
113. Barros, Decada I, part II, p. 99, Correa, *Lendas da India*, tomo I, p. 407.
114. Bulhão Pato, tomo II, p. 268.
115. The local tradition regarding the miraculous construction of the church at Quilon and the departure from there to the Coromandel coast is described by Castanheda, *História do Descobrimento & Conquista da India pelos Portugueses*, pp. 126-7.
116. Barros, Decada I, part 2, pp. 345-9; *Cronica de Descobrimento*, pp. 144-5; Damião de Gois, *op. cit.*, part II, pp. 24-7; *Cartas de Affonso de Albuquerque,* tomo II, p. 401.
117. Correa, *Lendas da India*, tomo 1, part I, p. 906.
118. Simão Botelho, *Tombo do Estado da India*, Lisboa, 1868, pp. 30-4.
119. *Ibid.*, pp. 34-5.
120. Here is the text of the agreement 'Diguo eu Eytor Rodriguez, feitor de coulão, que he verdade que eu concertey com a senhora Raynha de coulão secretamente, pro vertude de huum poder que pera yso tenho do senhor capitão moor e governador das Indias, pera que deixando – me ela ffzer hua casa de ffeytorya ffore no dito lugar de coulão, e sem peleja, nem guerra, nem outro nhum impedimento, que taal ffose em aue ouuese mortes d'omens, per que se a casa não fizesse, qeu eu lhe podese quitar da pimenta, que ela he obriguada a pagar a el Rey nosso senhor do asento da paaz, o que eu vyse que hera serviço do dito senhor, e per palavra me dise que eu quitase tudo o aque eu vise ser serviço

do dito senhor, que ele o conffirmarya a averya por bem; e por quanto a dita senhora Raynha quis antes dinheiro que outra cousa algua, por me parecer mais serviço do dito senhor concertey com ela na maneira seguinte- a saber- que leyxando-me ela ffazer a dita casa paçiffiquamente e come acima dito he, e dando pera iso toda a ajuda e ffauor por meu dinheiro, que eu lhe dese da ffazenda del Rey noso senhor duas mil Rajas, as quoaes leh fficava a paguar, por não ter dinheiro, em cobre e prata depois da dita cassa ser ffeyta, e eu metido nela he outro dia, lhe paguar as sobreditas duas mil Rajas, como sobre dito he, em prata: e porque isto tudo he asy verdade, e a dta senhora Raynha querer que lhe dises esta escrito por mim, sem outra pesoa d'iso saber parte, lho dey e ffiquo a conprir como nele he contheudo, comprindo sua Alteza as condições açima escritas: ffeyto em coulão a xxj dia de março de 519 anos' ref. Tomo do Estado da India por Simão Botelho', in Rodriguo José de Lima Felner, ed, *Subsidios para a Historia da India Portugueza,* Lisboa, 1868, p. 34; Gaspar Correa, *Lendas da India*, tomo II, p. 577.

121. Herman Gundert, *Kerala Palama*, Kottayam, 1868, pp. 140ff.
122. Simão Botelho, loc. cit., pp. 35-6.
123. Simão Botelho, *Tomo do Estado da India*, pp. 35-6.
124. Simão Botelho, loc. cit., pp. 36-9.
125. Correa, *Lendas da India*, pp. 906-7.
126. Couto, *Da Asia,* Decada V, part I, p. 3.
127. This was the first seminary built in India by the Portuguese. The seminary at Goa was established only in 1541. Panduronga S.S. Pissurlencar, *Regimentos das Fortalezas da India*, Bastora, 1951, pp. 229 ff.
128. 'Tombo do Estado da India por Simão Botelho', in Felner, Rodriguo José de Lima, ed., *Subsidios para a Historia da India Portugueza*, Lisboa, 1868, pp. 27-8.
129. M.S. Naravane, *The Heritage Sites of Maritime Maharashtra*, Mumbai, 2001, p. 54.
130. *Tombo de Dio* by Francisco Paes, 1592.

CHAPTER IV

Portuguese Shipbuilding in India

The Portuguese who came to India realized the need of installations for careening the ships in which they travelled from Portugal. They realized that the temperature of water in the Indian Ocean differed from that of the Atlantic. Besides, the amount of cargo carried in the vessels increased by leaps and bounds. They also faced stiff competition from Asian and other West European merchants. Naval confrontations became a common feature in the navigation in the Indian Ocean regions right from the first decade of the sixteenth century and especially with the appearance of the merchant marines of the Dutch and the English towards the beginning of the seventeenth century. Above all there was a great shortage of vessels and timber suitable for ships in Portugal. So the Portugese adminstrations became convinced that ships could be built in India making use of the local woods. They found that carpenters in India excelled in the selection of appropriate timber, being familiar with the building of ships for voyages on the high seas. Similarly, they came acros instruments of navigation employed by Indian mariners. It was against this background that the Portuguese started setting up installations in India for careening and building ships.They interacted with local ship builders directly. Therefore we may rightfully inquire into the details of the Portuguese naval architecture and look for the details of interactions.

The tonnage of the ships built by the Portuguese differed greatly in tune with the unprecedented and consistent increase in the volume of commodities exported from India to Portugal. The early vessels especially those under the command of Vasco da Gama for his first voyage were of minimal tonnage. The number and status of the passengers too increased during this period. The long duration of the voyage from Portugal to India and vice-versa necessitated the enhancement of the tonnage of the vessels. More amenities had to be provided in the ship taking into account the nature and number of the passengers. Similarly on account of naval encounters with competing west Europeans for a share in the maritime trade of India prompted the Portuguese to mount more cannons and other equipment on board. This had to be reflected in the volume and architecture of the vessels in the sixteenth century. The Portuguese used more Indian timber, especially teak, *angeli* and so on which were felled in the appropriate seasons according to the advice of local carpenters whose services were sought by the Portuguese. The masts were made of timber different from those used for the hull and other external parts.

The Portuguese set up their fortresses and naval establishments in places like Cochin, Cannanore, Calicut, Quilon and Cranganore on the Malabar coast. Cochin was an important centre of shipbuilding and repairs until 1663. Timber from the interior was brought in on the river Meenachil and through the backwaters to Cochin. Another route from where timber was brought to Cochin was the one via Periyar and Chetwai. Local carpenters were employed to select suitable trees and fell them in the appropriate season. Iron nails brought from Biscay in Spain were used on a large scale on account of the increase in the tonnage of the ships and also for strengthening the vessels against naval attacks in the seventeenth century. In caulking, the Portuguese seems, to have followed Indian methods. Elaborate narrations are found in the contemporary Portuguese writings of Fernando Oliveira, João Baptista Lavanha, John Huygen van Linschoten, the Dutch man and Ludovico di Varthema, the Italian.

The western coast of India has been famous for the best variety of timber for shipbuilding, especially teak wood. The southern extremity of the western coast, namely Malabar is known for the network of lagoons and small rivers connecting the hinterland with ports like Quilon, Cochin, Cranganore, Calicut and Cannanore. Therefore, from time immemorial, vessels were built in Beypore near Calicut, though not in Cochin which developed as a major port only subsequent to the flood in Periyar River in 1341 and the geophysical changes that took place at that time.

Meanwhile in Europe, since the time the Portuguese had obtained a copy of the map used by Marco Polo, they had worked on it and had made substantial progress in marine cartography and navigation. They developed suitable vessels for the India run based on the experience accumulated through years of maritime activities initiated by Henry the Navigator. The four ships in the fleet of Vasco da Gama that left for India from the river Tejo in 1497 did not exceed a tonnage between 100 and 200 as reported by the official chronicler of the Portuguese India, João de Barros. The Portuguese are credited with the distinction of being the pioneers in crossing the Cape of Good Hope and landing at a roadstead near Calicut on the western coast of India towards the end of the fifteenth century whereby direct links between the Iberian ports of the Atlantic Ocean and the ports of the Arabian Sea were established for the first time. The navigational techniques accessible to the Portuguese, coupled with the tenacious and adventurous will of Vasco da Gama, were to a large extent responsible for this feat.

The tonnage of ships and the number of decks increased in tune with the cargo taken from India to Portugal and the unprecedented growth of trade and movement of personnel. Though under Dom Manuel I (1495-1521) and Dom João III (1521-57) some of the ships were only of three decks, on account of the flourishing trade, the number of decks went up to six disregarding existing rules concerning the construction of ships. Dom Sebastião put an end to the lawlessness in the construction of ships and consequent shipwrecks. He was constrained to issue an order to limit the tonnage of ships of the India Run (*Carreira da India*) to 450. But after the death of Dom Sebastião, the tonnage of ships began to exceed the limit set by him. Again in 1621, the Portuguese king Philip asked the opinion of the board of experts on the subject of tonnage and the number of decks in a ship. In other words, the tonnage of the ship and naval architecture in general were always in the process of change during the sixteenth and seventeenth centuries in Portugal and Portuguese India.

Cochin on the Malabar coast had a royal shipbuilding complex (*Ribeira das Naus*) from the first decade of the sixteenth century. Later Calicut was chosen though for a short period by the Portuguese as a shipbuilding centre on account of the availability of timber in the hinterland and the possibility of obtaining the material easily. The Portuguese shipwrights and the theoreticians had written a lot about timber of various types that could be used for the construction of ships. They had been highly appreciative of the timber on the Malabar coast and the vast network of waterways that helped its transportation to the important centres of shipbuilding. The specifications prescribed in Lisbon from time to time were binding on the shipyards of Portuguese India in general and those on the Malabar coast in particular. Therefore, an attempt is made in this chapter to study the nature of timber used in shipbuilding and other aspects related to naval technology based on contemporary sources chiefly in Portuguese.

Just as the Portuguese were pioneers in using the maritime route via the Cape of Good Hope to western India and to ports on the Atlantic Ocean, they excelled, by and large, over the other European powers in building ships.[1] The writings of the Portuguese were scientific and systematic giving detailed sketches to scale for the building of ships. They realized the importance of the scientific construction of ships and so several works dealing with the systematic construction of ships were written. These are preserved in various archives in several parts of Europe.[2]

Various aspects of naval architecture are discussed in detail by the Portuguese authors of the period under study. They were very essential for building durable and steady seaworthy vessels on the Malabar coast under the Portuguese. Some of the authors like Fernando Oliveira had great experience in the shipbuilding centres of various parts of the world such as Spain, France, Italy, England and Arabia. He understood the crucial importance of scientific construction of ships. He writes: '. . . whereas a ship, even though built of good wood and well strongly fastened, will be useless, unless it is properly symmetrical. If it is lower than it should be, the sea will swamp it, if it is too high the wind will overcome it, if too narrow, it will be unable to carry sail, if too wide, it will steer badly, if one side is higher than the other, it will heel over to the great detriment of those who are in it.'[3]

Timber and its Qualities

The teak wood and similar ones that grew in the hinterland of Malabar attracted the Portuguese which prompted them to set up shipbuilding centres on the Malabar coast. All those who wrote about shipbuilding laid great stress on the wood.

Ocean-going vessels of the pre-industrial era were made chiefly of timber, the importance of which was recognized by people in the remote past like Anacharsis, the Scythian. He visited Athens *c.* 590 BC and is reported to have made an interesting and relevant statement regarding the crucial role played by timber planks in the construction of ships. He stated that the ship's side being four fingers thick, the passengers in the ship were only that far from death (i.e., four fingers). A Portuguese writer in the seventeenth century speaking of naval architecture quoted Anacharsis, the Scythian and wrote that nobody should take lightly the selection of suitable variety of wood for ships since the salvation of the sailors depended on a plank of insignificant thickness set between them and death.[4]

Depending on the exposure of the various parts of the vessel to water, wind and other forces of nature, different varieties of timber were chosen. No vessel was made of just one and the same type of timber.[5] Fernando Oliveira, writing in the sixteenth century, speaks of this lucidly. He compares the ocean-going vessel with the body of an animal. The skeleton of the body can be likened to the frames of ships because it supports, strengthens and gives shape to the body. The frames of the ship do the same in the hull. The planking of the ship is compared to the skin in animals. The hull of a ship requires strong and hard timber. It has to bear all the weight of the ship and to withstand the forces of the sea and wind. The planking on the other hand, must have flexibility a softness allowing it to be bent and joined to the frame according to the curves of the ship's side.[6] In general, wood for the construction of ocean-going vessels should be tough, dry, of bitter and resinous sap. It should be tough and strong to withstand the impact of sea and wind. It has to be dry from dampness of the waters before the ship is conserved in the water. If only the sap of the tree is resinous, it can get itself rid of water. Bitter sap can keep off the shipworms. Pliability is required for the timber in bending and joining as mentioned in the case of planking above.[7]

The Portuguese wrote that teak (*Teca*) and *angeli* (*andira*) of the Malabar coast were the only trees that had all the above-mentioned qualities so that nature seemed to have created them exclusively for naval architecture.[8] This was the chief factor that impelled them to start shipbuilding on the Malabar coast.

Though Portugal had other sorts of timber suited for the small vessels in the period of discoveries, it was rather difficult to find trees capable of producing timber in the size required for the vessels of the India run (*Carreira da India*) in the later sixteenth century, when the volume of shipment grew beyond expectation. So attempts were made to establish shipyards in various parts of coastal Malabar where there was a large supply of angely and teak that provided long and sturdy pieces of timber. Moreover, the nature and the temperature of water in the Indian Ocean required timber different from that which was used for ships plying in the colder waters of the Atlantic and Europe in general. Shipworms were born either inside the timber or entered into the timber from outside in the waters of the hot climate. Therefore, Fernando Oliveira writing in the sixteenth century recommended and approved the selection of timber for ships according to the climate of the area where they were to ply. He adds that the ships should be built with the timber of the land they were to visit. This would mean that ships plying to and fro in the Indian Ocean regions should be constructed with the timber available in these regions. In fact, all these explain the interest shown by the Portuguese to get shipyards established in India and to obtain timber of *angeli* and teak from the Malabar coast and other areas of costal India.

The required number of suitable trees was cut during the summer from the hill regions of Malabar, sometimes under the supervision of Portuguese officials. Logs were dragged to the riverside by elephants so that during the monsoon season they could be floated on the water and brought to the shipyards, as described by one of the Portuguese master carpenters in the Cochin shipyard, without cost. The Portuguese did not pay any taxes for timber to any body.

João Anes who was in Cochin for over fifty years from 1502 and was the master carpenter of the Portuguese shipyard in Cochin wrote about the way in which timber from the hilly regions in Malabar was brought to Cochin by river. He wrote that two elephants were kept in the summer season in areas where trees were cut. With the help of these elephants the logs were brought to the places of shipbuilding without any customs duties to be paid to anyone.[9] It was reported on 10 November 1518 from Cochin by João Anes, that two *galeotas* and one *galeão* were built in Ponnani and two *fustas* in Paliporto near Cranganore.These were much better than those built at Calicut.[10]

A number of rivers flow from the hills of Kerala to the big lagoons, which again connect these hilly regions to the port of Cochin. It was through these rivers that the Portuguese got the required timber to the shipyards from the interior places as described in the Portuguese reports of the sixteenth century. A lot of *angeli* timber was brought to Cochin during the monsoon through the Cranganore River which touches Chetwai, north of Cranganore. The land through which the river flows belonged to the ruler of Udayamperur. Another river connecting the island called Chembu went further into the interior. Chembu belonged to the ruler of Vadakankur. There was a place called Velloor near this river and a lot of merchants living in that place dealt in timber. They brought required timber through this river to the Cochin shipyard.

To the south of Vaduthala there was Pallipuram, belonging to the king of Vadakankur. Vaikom and Thalayalam were also in the same kingdom. A lot of timber used to pass through Thalayalam to Cochin on the river. This river touching Vechur, Kudavechur, and Kudamaloor, originated from the hill ranges from where large volume of teak and *angeli* was brought to Cochin. The river touching Ponnani connected places in the interior. Cochin received masts for ships and teak wood from there. There is still another river north of Ponnani passing through the lands of the *Kaimals* to the hilly regions which belonged to the king of Calicut, the Zamorin. Timber was also brought to Cochin though this river. The river joining Cranganore and flowing through Alengatt and Parur brought a lot of teak and masts of *angeli* to the Cochin shipyard regularly since the rulers of these regions were friends of the Portuguese. *Angeli* timber was also brought to Cochin through Parur in the north. But the best variety of timber that was brought to Cochin came from the east of Parur that also flowed through the land of the Vadakankur raja. The river passing through Anapamparao (Anapamparambu) used to bring a lot of timber from the interior places to Cochin.[11] Thus we find that Cochin was supplied with plenty of the best sort of timber for shipbuilding. The anonymous author of *Lembranças* speaks of a large number of vessels available in India some of which were constructed in Cochin.[12]

The supply of the best sort of timber for shipbuilding encouraged the Portuguese to open their shipbuilding centre at Calicut in the period after 1513, though it did not survive long. A lot of teak wood was available in the hinterland of Calicut especially in the present Nilambur forest from where through the river it could be brought to Beypore renowned for shipbuilding from early times. Duarte Barbosa was in-charge of shipbuilding at Calicut. The Portuguese got two vessels built at Calicut in 1515 under his supervision for the Muslims of Mocha. Afonso de Albuquerque, the then Portuguese governor, sanctioned the payment of a certain amount of money to Barbosa.[13]

The Portuguese who valued greatly the timber available in India took the good varieties to Lisbon for the construction of ships. Apart from the timber available in India, the Portuguese shipwrights had to depend on that available locally in Portugal to build ships to be sent to Indian waters. Portuguese experts writing in the sixteenth and early seventeenth centuries on naval architecture and selection of good varieties of timber, refer frequently to the scholarly suggestion given by Marcus Vitruvius Pollio (*c*.90-*c*.20 BC), Roman engineer, scholar and builder who wrote *De Architectura* in ten volumes. Taking into account the need for hard timber for frames and timber for planking, wooden materials were selected. The Portuguese carpenters coming over to India were asked to find out suitable timber specified in Lisbon for various parts of the ship. But these were not easily available in India.[14] Whereas in Portugal, they were asked to work on cork, oak (*Pinheiro manso, Quercus suber*) for the frames (Hull), and on stone pine (*Pinho manso, Pinus pinea*) for planking.

The carpenters engaged in shipbuilding in Portugal and Portuguese India were advised by contemporary naval architects to give greater attention to the specific quality of the timber used in shipbuilding rather than relying on the species of the tree. They were supposed to select strong wood for frames. For planking, the timber should be flexible, solid and close-grained below the water, light in the upper works and long, straight and clean for masts. For all these it was recommended that the tree should be healthy and free of knots, which caused the Latins to call it 'spotty wood'.[15] The best results were obtained when the trees used for shipbuilding were neither very young nor very old. The timber of the young ones was not strong and that of the old trees was generally affected by woodworm and liable to rot. The naval architects of the period recommended the use of different types of timber for different parts of the ship. In other words, a ship in Portuguese India was not made only of one kind of timber only.[16] They attached great importance to the climatic condition of the area in which a ship was to ply. Moreover, the qualities suitable for longevity and type of sea water into which the ship was to be launched were the prime consideration rather than the species of timber since the quality of timber differed from place to place and from site to site, regardless of the species. Hence, Portuguese carpenters coming to the Malabar coast were supposed to take into confidence the people of the place where the ship was to be built to find out the best variety of timber suitable for the construction. The Portuguese shipwrights in India took the assistance of a number of local carpenters. The Portuguese master-carpenter in Cochin wrote to the king of Portugal about the availability of good carpenters in places like Cochin, and discouraged sending Portuguese carpenters to India.[17]

The desired advantage of the best type of timber for shipbuilding will not be obtained if the tree is not mature and even if it is mature if it is not cut in the right season. Hence the Portuguese writers of the sixteenth and seventeenth centuries instructed the persons connected with shipbuilding that the trees should be cut only when they were ripe. However, they added that the trees should not be very old for fear of woodworms. If they were not mature, the timber would be soft and could not be used for shipbuilding. Timber of immature trees caused shrinking and opening up joints and created changes in the work as well. It was further indicated that the time for maturity differed from tree to tree according to region and even within the same species difference is noticed according to region and sites. Maturity is obtained faster in a hot climate than in a cold one.[18]

As indicated above, even if the tree is mature, if not felled in proper season, the timber is likely to be spoiled. The Portuguese advised on a stipulated season. They differentiated the seasons in India and Portugal. Quoting Vitruvius, Fernando Oliveira said that trees in summer in Europe were like women who became pregnant and gave birth and raised their offspring, and then became less strong, for they passed on the great part of their sustenance to the young ones, both in the womb and outside and in the same way, the trees passed on their sustenance to the fruits and thus became weak and constituted the source of new wood as well.[19] Trees in Portugal should be cut only during the period of two months, namely one month before the beginning of winter, 24 December and one month after the beginning of winter. During this season, the sun is quite far away from the trees putting them to rest and making them recuperate all the energy and strength spent during the active life in spring and summer. He added that it is the sun that made trees gives birth to fruit. By autumn the trees began to take rest after these activities. So, the best season for felling in Portugal for shipbuilding, as far the effect of the sun was concerned, was the one suggested above.

The major part of India being in the tropical region, it has longer summer. Oliveira, therefore, advised that trees be felled during the solstice. He said that *angely* and teak trees being closer to the sun, they got a lot of sustenance with the help of the sun. Therefore they were always strong.[20] They could be cut when the fruits were ripe. Great care was to be taken in cutting the trees meant for shipbuilding and the naval architect should employ more diligence than the civil architect. As suggested by the contemporary Portuguese writers, the weakness of timber used in the construction of a house was of little consideration as compared to that of a planked ship. Realizing the importance of this, they went further into the details of the moon. Since the moon is closer to the earth than any other planet, it feels the impact of the earth faster, as does the earth. Just as the distance of the sun affects the trees, that of the moon also affects them.[21] During the waning of the moon, the dampness in the trees may be greatly reduced and the trees become dry. Therefore woodworm does not affect them. The general suggestion therefore, was to cut the trees for shipbuilding during the time when the moon was on the wane within the period prescribed, taking into account the condition of the tropics.[22]

The contemporary Portuguese writers suggested that it was better to leave the trees half-cut for three or four days to permit the dampness to disappear and drain out. Further they insisted that the felled trees should be left many days in the field or in the shipyard or in the salt water. If the timber was immediately worked on, it might shrink while drying and may open up cracks that are greatly detrimental to shipbuilding. The dried timber should again be soaked before work is started on it.[23]

Thus, the Portuguese naval architects took extraordinarily diligence in the preparation of timber for shipbuilding since a small plank in a ship was of immense importance and the life of the passengers and crew was always in danger if the plank was not good.

Stages of Shipbuilding

The different materials to be made ready for building diverse types of ships, the measurement of the various parts of the ships, the way in which they had to be joined together, caulking and other details of maintenance of the ships are given in detail even with specific design, by Portuguese writers of the sixteenth and seventeenth centuries. The earliest written prescriptions are from Fernando Oliveira. Similarly, João Baptista Lavanha and Manuel Fernandes provide the details of the entire operation of shipbuilding emphasize the stages from the felling of the trees, seasoning of the timber, design of the keel, stem, sternpost, master frame, *Braços*, *aposturas*, transom and fashion pieces besides the drawing of moulds.[24]

Use of Nails

Many of the early sixteenth century reports regarding the vessels used in the Arabian Sea belonging to the Indians lay stress on the use of coir for joining the planks. Indians used iron nails sparingly because the ships could easily sink when the ship with iron nails came in the magnetic zone found in the rocks under water. Many writers noted with great surprise that, unlike the Europeans, the Indians did not use iron nails in the construction of ships in India.[25] Ludovico di Varthema of Bologna who spent some time in Calicut gives us a report about the vessels built there before the Portuguese opened their shipbuilding centre. According to his narration, vessels of three hundred to four hundred butts were built in Calicut. Though carpenters did not apply oakum between the plank. The planks were so well joined that water was repelled in the most excellent way. Pitch was used outside. Cotton sails were used for ships in Calicut. At the foot of the sails they used to carry another sail with a view to catching more wind. Varthema holds the view that the use of a second sail in India was totally different from the practice of the Italians who used only one sail. The anchors used by the people of Calicut were marble pieces of 8 palms' length and 2 palms breadth attached to two large ropes. He praises the quality and volume of timber for shipbuilding available in Calicut and points out that his country did not have as good a variety of timber as could be found in Calicut. It may be mentioned that these ocean-going vessels were not very small. To cite an example, we have reports in Portuguese about a ship owned by two merchants of Cochin namely, Mammale Marakkar and Cherinina Marakkar, which was approximately of 600 tons and carried seven elephants and over 300 men on board. This naturally was far superior to the Portuguese ships of the time and was an enormous ship in the context of the time.[26]

The Portuguese writings in the early part of the sixteenth century lay stress on the marvellous victories courted by the Portuguese sea-men over Indian ships and attribute such victories chiefly to the weakness of Indian ships. With a small number of Portuguese vessels and mariners, a large number of Indian ships were destroyed or captured along with the crew. They specifically mention that since the Indian vessels were not built with iron nails, it was easy for Europeans to overpower the coir-sewn vessels of the Malabar coast. Such vessels were not able to withstand the impact of cannon shots fired from the Portuguese ships.[27] The joins of planks by coir and even by wooden nails gave way on the impact of cannon fired from the Portuguese ships. It was also possible that by using wooden nails, if at all they were used, the ships were rendered weaker than those made with iron nails.

The Portuguese make mention of wooden nails used in France, Holland, Zeeland and England for building ships. These nails had the advantage of not rusting. But these architects insisted on the exclusion of wooden nails for the ships plying in the Indian Ocean regions for a couple of reasons. First, the wooden nails had to be used more closely than the iron nails and had to be thicker than the latter. The use of bigger nails and more wooden nails in a ship weakened its body. The timber being closely bored for wooden nails naturally becomes frail. The vessels that were used in the Indian Ocean regions had to be in water of a higher temperature than that in European waters. The temperature of the Indian Ocean is more conducive to the growth of woodworms than that of the Atlantic. Therefore, the wooden nails could easily be attacked by these worms and the ships could be in danger in a short span of its life time. Hence wooden nails were not recommended for ships in the Indian Ocean regions. Another important factor that prevented the Portuguese from using wooden nails in the construction of ships in India and Portugal for the India runs was the size of the timber used in vessels of higher tonnage. As trade and the movement of the people between Portugal and India grew unprecedentedly, the tonnage of ships went up from 150 to 600, 800, 1,000 and even 1,200. Hence the timber used in such ships had to be thicker than the one used in early days of discovery. Thus wooden nails, if at all they were used, had to be thicker and longer in proportion to the pieces of timber for construction. This would entail close boring which in its turn weakened the structure of the ship. Therefore wooden nails could not be used in such vessels if they were to be durable and strong.[28] The other alternative was to use nails made of copper or iron. Oxidation could not consume copper nails as quickly as iron nails nor could moisture corrupt copper nails as it could do with iron nails. Hiero Syracuse is said to have built a ship using copper nails. This was very lasting and well-known. But copper nails were expensive. Therefore the best solution suggested by the architects for the construction of ships in India was to use iron nails.[29]

Iron nails should be well-tempered, strong and well-cast. The best variety of iron nails was imported into Portugal from Biscay in Spain and from there to India. The people in Biscay knew how to make well-tempered iron nails, which would not break during driving and riveting. The iron nails made in Lisbon were not as good as the ones available in Biscay. Thus iron nails that were generally used in the Indo-Portuguese shipbuilding centres like Cochin were those brought from Biscay via Lisbon. The master-carpenter in Cochin shipyard used to write directly to the king of Portugal to ensure that iron nails and iron in sheets were sent from Portugal to Cochin as they were cheaper and of better quality than the ones available in India.[30]

Discussing the reasons why iron nails were brought from Portugal to India for shipbuilding, a sixteenth century document presents two points. The quality of iron available in India was much inferior to the one brought from Europe. Chaul, Bassein and the kingdom of Vijayanagar used to provide iron. The iron from Bassein and Chaul was the best available in India, but even that was much below the quality of iron imported from Portugal. Besides, the iron from India while being worked on had a lot of loss. The people working on iron in India were not as expert and hardworking as those of Europe either. Apart from the quality of iron nails, the price also counted much. The price of iron after all its loss was lower than that of India. A quintal of nails used to cost 3 *cruzados* in Portugal while that of India cost 8-10 *pardaos.*[31] Iron as such cost 600 *reales* per quintal in Portugal and it was available in plenty. The contractors of Belgium supplied this to Lisbon. Even if 1 quintal of iron was available for 450 *reales* in India after the waste while being worked on, it would cost the same as in Portugal. But the quality was still inferior. So, it was recommended that the iron materials for shipbuilding in Portuguese India were to be imported from Portugal.[32] In order that the iron nails driven in the timber might permanently avoid entry of water, these were to be caulked. A number of items were used for caulking.[33]

Tonnage of Ships in Portuguese India

The *naus* of the Portuguese in India, especially those involved in trade between the Malabar coast and Portugal underwent a drastic change from the time Vasco da Gama reached India in 1498. The tonnage of the ships in the first voyage of Vasco da Gama was between 50 and 150 or maximum 200.[34] The tonnage of a ship of an Indian merchant captured by Cabral in 1502 near Ponnani definitely was much higher than the Portuguese caravel under the admiral, Vasco da Gama, in his first voyage , because it carried seven elephants and 300 passengers besides victuals. This was a vessel owned by merchants of Cochin and had a tonnage of approximately 600. The ships that plied in the Arabian Sea with cargo meant for distant places like the Red Sea or Persian Gulf regions in the west or Malacca in the east were of great size. In other words, the cargo ships for long-distance trade were always huge, to justify the expenses involved in distances to be covered and the risks to be faced.

As soon as their trade and transportation of personnel began to pick up, the Portuguese saw the the need for vessels of significant tonnage. They launched the construction of large ships of different ranges of tonnage. The mounting of heavy artillery also compelled them to have large vessels. They realized the significance of the resistance put up by the Zamorin of Calicut and Malik Ayaz of Diu in collaboration with the Arabs supported by the Venetians in 1508 and 1509 at Chaul and Diu respectively. Hence, after the discovery of the direct sea-route and the establishment of commercial relations the tonnage of the ships in India rose. The average tonnage of these ships under Dom Manuel and João III was between 500 to 1000. The contemporary naval architect, Fernando Oliveira states that there were several vessels of between 800 and 1,000 tons, and these made always the best and the safest voyages.[35] But it seems that the tonnage of the ships of the Carreira da India was higher and that there were even ships with seven decks. On account of the great abuse and shipwrecks caused probably by overloading in violation of all the existing instructions rather than because of the higher tonnage, King Sebastian was constrained to reduce the tonnage to 450 but after his death the tonnage of the ships in Portuguese India rose. This situation prevailed till 1621. A board was set up in 1622 to go into the details of the tonnage and opinions were collected from concerned persons.[36] Even in the 1570s as reported by Fernando Oliveira, there were a few miserly people who argued, probably on account of the large investment needed for bigger vessels, for ships of lower tonnage. He very scientifically and convincingly argued for ships of higher tonnage for long distance voyages to Portuguese India.

The proposals to have higher tonnage for the vessels of Portuguese India were defended logically in the second half of the sixteenth century. Transportation of large volume of cargo required vessels of great tonnage. Vessels of bigger size were not easily liable to sink while narrow ships were quicker to submerge. This was demonstrated by an example. If a plank is thrown flat upon the water, it will remain afloat. If on the contrary, the same plank is thrown on water on its edge it will immediately go down in the water until it remains in the same position, even though the weight of the plank is the same. The different reactions seen here are on account of the resistance of the water that it holds beneath itself. When the ship is wide it has a lot of water under it and the resistance of the water will keep the vessel floating. A narrow ship will not have as much water under it for resisting its submersion as a wide ship. The same point can be clarified from another point of view. This is in the context of equipoise or matching the weight, which is not clearly distinct from the earlier point. The water beneath the wide plank or ship weighs more than the plank ship. A heavier object will keep less heavy object above in case of a liquid substance. Thus the water under big ships will prevent them from sinking. Air being lighter than water, the wide-bodied ships containing a large volume of air will be kept floating on the water. The water draws such beamy ships upwards and provides them resistance from sinking. The beamier the ships are, the more air they contain. Therefore, wide-bodied ships in proper proportion are safer than narrow-boded ships of the same proportion.

The long voyages undertaken by the ships of Portuguese India required large volumes of victuals. If the ships were small they would be filled with mariners and their victuals leaving no space for merchandise. Fernando Oliveira argues out his case for large vessels against the proponents of small ships for the India run. He states that if the tonnage of the ships was less, the expenditure would exceed the gain. Small ships were less safe than the larger ones for two reasons. First, more men and armaments could be accommodated in big vessels and they could be used against robbers both on sea and in the ports they visited. Second, the very sight of larger vessels could terrorize an enemy. Countering the arguments of those who would say that when larger vessels would be lost the loss would be greater, he says that only in very rare cases were large ships lost whereas the case to the contrary was much more prevalent. So, he very strongly upheld the idea that the tonnage of the ships should be enhanced and advised the Portuguese to follow the tradition prevalent during the period of Dom Manuel and D. João III.

Oliveira confirmed his idea of building ships with higher tonnage by citing the practices of antiquity. The grain ship of Ptolemy Philopator was able to carry 400 sailors and 3,000 fighting men. Similarly, Hiero, the king of Sicily, had larger ships of more than 1,000 tons. Hence, even by citing cases from the remote past he buttressed his argument that Portuguese India should build vessels of greater tonnage for long distance voyages. Fernando Oliveira seems to have influenced the shipbuilders of Portuguese India greatly. It was suggested by the Portuguese admiral, João Corte Real who was in the service of the Portuguese king for sixteen years that the ships should have four decks, not three, so that more cargo and personnel could be transported. This could help in naval battles while facing other ships.[37] This opinion was submitted to the king according to the ideas of the officials of the shipyard and Admiral Corte Real. The argument of Admiral Corte Real was submitted to the king on 1 January 1622. The document is entitled 'Discurso que fez sobre as naus da Carreira da India' and is found in maço 347 of the collection *Conselho Ultramarino de Biblioteca Nacional de Lisboa*. But another trend began to gain popularity against such massive structure of ships, which prompted the then king to refer the matter to a committee of experts. A letter issued on 22 January 1622 resolved that there should be only three decks for the ships involved in trade in Portuguese India. Even after giving the order on 22 January 1622 the king conducted further inquiries and sought opinions of several individuals and committees regarding the size of *naus*. The long discussion is contained in more than forty-two printed pages which show the importance given by the king to the matter.

Types of Ships Made by the Portuguese in India

A few commonly used vessels are mentioned below though there were still many more.

ALBETOÇ

It was a vessel of Indian origin which the Portuguese used for pleasure trips. It was also employed in naval combat. It could carry 20 passengers on board.

ALMADIA

This was of Indian origin with a length of 80 feet and breadth of 6 to 7 feet with pointed ends. It is operated by sails and oars at a high speed. It was used for transport of victuals and as a relief vessel.

BALANÇO

This was of Asian origin and was employed by the Portuguese for the service of large ships of fleets. This was propelled by spade-oars. It could carry 10 to 12 persons on board.

BALANDRA

This was used for coastal navigation. It had a deck and the large *balandras* were of 80 tons and were provided with a crew of three or four sailors.

BATÃO

It was small and light vessel propelled by oars. It looked like an *almadia*. It carried 12 gunmen besides the crew.

BARCA

This is of Portuguese origin. It was propelled by hackle oars (*remo de espadela*). It had a tonnage of 20 to 25 with a single deck. Usually it had 14 oars and 2 masts.

BERGANTINE

It was provided with 8 to 10 benches for oarsmen and equipped with one sail. It was armed with 12 to 20 artillery pieces, two lateen masts and 1 deck.

BOTE

It was a vessel of oars much smaller than the life-boats used for carrying light cargo especially pepper to the factories.

CARAVEL

Caravel was of Portuguese origin. It was usually of 200 tons, provided with two or three masts which were exclusively lateen rigged. The Portuguese caravel was basically a round ship.

CARRACK

It was a merchant vessel of considerable size and was the largest round ship sailing in the Indian Ocean regions. A minimum depth of 60 feet of water was required to float the ship. It had more than three decks and was capable of carrying 2,000 persons on board.

The stern and the prow were higher than the upper deck. There was a two storied platform on veranda between the two castles.

CATUR

It was a small ship of oars like Brigantine used for fishing and piracy. It was furnished with pointed bows and arrows. It had one mast.

FRAGATA

It was a sailing ship without castles, smaller and lighter than the *nau* but with two decks and 30 to 60 pieces of artillery mounted. The ships of this type sailed always in squadron and discharged important duty in naval battles. It had a mast of good height inclined towards the stern where there was one large lateen quadrangular sail with pulley.

FUSTA

It was very commonly used by Muslim traders. It belonged to the family of the oar ship. It had often 35 oars operated by 37 oarsmen.

GALIOTA

It was a small boat of complicated structure used for combats. It had 15 to 20 benches.

GALLEY

It was a battle ship of small tonnage with triangular lateen sail. It carried 25 to 30 oars on each side with 3 men for each bench.

GALLEON

It had two decks with a prow. The number of sails was not definite. The bigger Galleons had 4 masts, 2 round in the front and 2 lateen at the rear. These were the foresail masts, main masts, mizzen mast and *contra artimão*.

MANCHUA

It was used by the Portuguese in Goa. It was a small oar ship used on the Indian coast and equipped with one quadrangular sail. It could carry 4 to 9 pieces of artillery.

NAU

The term '*nau*' had not been used with any specific naval meaning. It was originally a generic term for *navio*, a high sea ship. It was meant for trade. The early *naus* were between 100 and 120 tons. They had two decks, the first one extending from the rear to the front containing the cargo hold, store room for water and provisions, cables, clothes, gun-powder, etc. The second deck at the prow had at the rear the captain's quarter-deck covering the castle of *bombardeiros*. By the second half of the sixteenth century *naus* with three masts appeared. Since there was not great difference between warship and merchant ship, naval experts suggested the use of the *nau* for India run. *Naus* between 500 to 800 tons began to be common in the India run.

PARAU

It was like a *Fusta* with 18 to 20 benches without deck. Besides oarsmen, it could carry 120 to 130 men of arms and 32 bombards.

PATTEMARIM

It was a light ship of 5 to 12 tons used for coastal navigation in India. It was equipped with two masts.

PINACA

It was a light and narrow vessel operated with oars and sails and three masts with a square stern.

SAMBUCO

It was a small ship of flat bottom without deck. It was like small *Galley* manned on each side by 20 to 30 oarsmen. It could carry three to four pieces of artillery and more than 100 archers.

TONE

This was used for fluvial navigation and was employed for taking pepper to the factories. It was a flat bottom ship without deck and with single mast and oar.

URCA

It was a cargo vessel of low speed with flat ribs, broad on the planks and round in the rear. It carried two masts.

Centres of Shipbuilding on the Malabar Coast

The Malabar coast, as indicated above, had a number of centres where ships could be built. The availability of timber and the water transportation prompted the Portuguese to start shipbuilding on a firm footing there. But the political atmosphere too played an important role. Even though Calicut and especially Beypore was the most important centre of shipbuilding for centuries, the Portuguese shipbuilding activities could not thrive there on account of the hostile attitude of the Zamorin and his Muslim lieutenants. But Cochin, which turned out to be a good natural port after the flood in the Periyar river, was able to provide the most congenial centre for shipbuilding. This was further enhanced by the amicable attitude of the king of Cochin.

Cochin

Dom Manuel, while sending Francisco de Almeida as the first viceroy to India in 1505, instructed him to start a shipbuilding centre in Cochin on account of the availability of good timber and the geographical location of Cochin, and to build oared ships.[38] It is reported that the mast used in the ship *São Gabriel* belonging to the first voyage of Vasco da Gama was dismantled at Cochin in 1506 for building another vessel there.[39] Around 1509 even private persons were involved in building ships in the vicinity of Vypin since a large supply of timber was available there. A certain Jorge Barreto, captain of the Cochin fortress, obtained timber from the king of Cochin to build ships for himself rather than for the king. João Anes, the master carpenter of the royal shipyard, testified to the fact that timber was obtained for the personal use of Barreto.[40] The naval establishments in Cochin were used for equipping and repairing the necessary number of vessels that were sent to Goa in 1510 for its conquest by Afonso de Albuquerque.[41] The ship called *Santa Catharina de Monte Sinai* of 800 tons (10,800 quintals) was built in Cochin under the orders given by the then Governor Afonso de Albuquerque. This ship needed water for a ship of 200 tons only. This was constructed between 1510 and 1512.[42] Another ship was built in the Cochin shipyard in 1512.[43] The ship *S.Catharina de Monte Sinai* was sent to Lisbon in 1518 loaded with commodities.[44]

By October 1514 three caravels were built in Cochin and the construction of two *Galeotas* was undertaken in the same year.[45] Several vessels like *Galeãos*, *Galees*, very small vessels, as *Nau*, *Santa Cruz* and *Galeota* were built in Cochin in 1527.[46] The ship *Santa Cruz* was taken by João de Sepulveda as the flagship to Lisbon in 1545.[47] There was a host of people of various departments of the shipyard in Cochin towards 1554 such as master of the shipyard, master of caulking, master of the blacksmiths, master of cordage, and master of cooperage.[48] The viceroys and governors of Portuguese India took great care in the development of the shipyard in Cochin. The governor Francisco Barreto in 1558 issued an order prohibiting the transport of timber from Cochin. In 1572, Viceroy Dom Antonio de Noronha issued another order setting apart the 'one per cent' duty levied from the port of Cochin for shipbuilding at Cochin. In 1573 the viceroy asked the municipality of Cochin to elect two officials, one for supervising the construction of ships and another for the collection of the 'one per cent' duty from the port to help in the shipbuilding activities.[49] Mathias de Albuquerque, the viceroy in India reported to the king of Portugal that Cochin was an ideal centre for shipbuilding for Portugal. King Philip ordered that at least two large vessels be built in Cochin every year.[50] Even at a later date, Aires de Saldanha, the viceroy was asked by the king of Portugal to encourage the shipbuilding activities in Cochin.[51] From the beginning when Portuguese contact had been established the king of Cochin helped the Portuguese to get necessary wood. The timber available in Cochin, especially teakwood, was reported to be far superior to that available in Bassein. For this reason shipbuilding activities continued to progress well in Cochin under the Portuguese.

João Anes as mentioned earlier was the master carpenter in the shipyard at Cochin. He worked there from 1519 to 1557 in this position. However, he was on the Malabar coast from the time of Afonso de Albuquerque and Lopo Soares. The latter appointed him as the master carpenter in 1519.[52]

Calicut and Cannanore

Calicut for centuries had been known as a great centre of shipbuilding and navigation. The Portuguese had established a factory at Calicut in 1500. It had a short life. So, no attempt was done to start shipbuilding at Calicut in the first decade of the sixteenth century. But as soon as the fort was established agreement of peace and friendship signed on 24 December 1513, Afonso de Albuquerque issued orders to build a *Galley* in Calicut in 1514.[53] The King of Calicut himself suggested to Afonso de Albuquerque that on account of the abundant supply of all varieties timber at cheaper rate at Chaliyam, the Portuguese could start building ships there.[54]

They took steps to take advantage of the facilities at Calicut, especially the availability of the best variety of timber. Afonso de Albuquerque in September 1515 ordered two ships to be built at the Portuguese shipyard at Calicut at the cost of Chetti merchants of Calicut under the supervision of Duarte Barbosa who was appointed as factor to deal with the construction of these ships.[55] In the same year, Duarte Barbosa got constructed two *Galleys* for the Muslim merchants of Mocha. This was done when Francisco Nogueira was the captain of the fortress and Gonçallo Mendez the factor at Calicut. Thus the Portuguese installations at Calicut since 1514 built ships both for the Portuguese and others under their supervision. We find that *Galleys* were the ones constructed here. This must be on account of the availability of better timber in abundance at lower rates.

Comparatively timber for shipbuilding at Chaliyam near Calicut was cheap. Another ship was made ready by October of the same year.[56] Occasionally shipbuilding took place also at Cannanore. A caravel was made ready there in 1514.[57] The Portuguese could not go ahead wih shipbuilding in Calicut for long. By 1525 their fortress was demolished. Hence, the Portuguese continued building ships chiefly in Cochin and occasionally in Cannanore. Though a new fortress was established at Chaliyam in 1531, we have no details of shipbuilding there.

Shipbuilding in Goa

When Afonso de Albuquerque conquered Goa in 1510, he was surprised to find a number of ships, naval spare parts and artillery pieces of different calibre there. He praised the expertise and excellence of the local carpenters in a letter written to the king in 1510.

Towards the second half of the sixteenth century Lisbon was

convinced of the fact that the ships for the India-run could be built in Goa. Even towards the last quarter of century the king wrote to the Viceroy Dom Duarte de Meneses on 22 February 1585 that ships of 500 to 600 tonnage could be built in Goa on contract as in Cochin and northern parts. He mentioned that from the information he had received and from experience, it was learnt that building ships in India would be cheaper and faster in comparison to shipbuilding in Portugal. Moreover there was shortage of timber in Portugal for shipbuilding. The ships built in India were stronger.[58] Again on 3 March 1594 the king wrote to the Viceroy Dom Mathias de Albuquerque that on account of the shortage of ships and timber in Lisbon, ships of 500 to 550 tonnage should be built, purchased, or built on contract in Goa. Since the ruler of Cochin had plenty of timber for shipbuilding, his services could be made use of, especially since he had more workers and expenses would be much less than in Lisbon. The teak wood available in Cochin was better than that of Bassein. He urged the viceroy to take it in good earnestness since the matter was of great importance.[59]

Goa shipyard produced ships of great fame. The *Nau Cinco Chagas* was constructed by Viceroy D. Constantino de Bragança in 1559-60. She served in the *Carreira da India* for twenty-five years and made nine to ten round-trips apart from other voyages. She was the flag ship of five viceroys before ending her days in Lisbon. Another well-known ship built in Goa was *Galeão Bom Jesus.* It was considered one of the 'noblest vessels made in India. Another ship of fame was the *Nau Madre de Deus*, a huge ship with three closed decks, seven storeys a main orlop, a forecastle and a spar deck of two floors. It measured 163 feet from beak to stern and about 47 feet across the second close-deck.

Other Portuguese shipbuilding centres were Bassein and to some extent Daman. Bassein was the main shipyard of the northern province of Portuguese India. The areas around Bassein yielded timber of the best quality.

Shipbuilding centres in Bassein were some of the oldest of the area. Even before the Portuguese occupied it in 1534, it was known for shipbuilding. Agashi had a site on the Sopara creek with an outlet into the Vaitarna estuary. It used to be a timber mart exporting timber to the Gulf countries. In 1540 the king of Portugal got a large ship of 700 tonnes built at Agashi. This was used for six years in the *Carreira da India*. There were other centres in Portuguese Bassein for building ships. Papdi, on the creek leading to it from the Ulhas estuary, Naugaon and Koliwada on the right bank of the Ulhas River beyond the quay to the east of the sea-gate of Bassein fort served as ship-building centres. Craftsmen were recruited from Sutarpada lying in between, and Bahadurpura, the Muslim settlement predating Portuguese Bassein. The Portuguese had brought in master shipwrights from Portugal, but the working class carpenters were Hindus of Agashi and Koli castes and Christian converts. Rui Leitão Veigas and his son Fernão were master builders of Bassein in its heyday. François Pyrard of Laval while speaking of Portuguese ships writes:

> I have heard it said by the Portuguese that no vessel ever made so many voyages from Portugal to Indies as a certain carrack that was built in Bassains (Bassein) which is between Goa and Cambaye. It made as many as six. Those built in Portugal ordinarily make but two or three at most, but the majority make but one. This place, Bassains is to the Indies what Biscay in Spain is here, for all the vessels built for the kings of Spain in the Indies are constructed there, because no country yields so much timber.[60]

The timber hinterland for shipbuilding in Bassein in the sixteenth century was Tungar and Bassein hills. Later the area extended to Vaitarna-Tansa valley upto Wada and Jawhar. Timber cut in these forests used to be floated down the Vaitarna and its main tributary to the estuary, to be collected at Arnala, Agashi and Papdi. The Portuguese set up a major timber mart at Manor and fortified it to regulate and organize the timber flow from the Vaitarna. The quality of teak available in these forests was very high. Relative cheapness in construction was another factor for the development of Portuguese shipbuilding in Bassein.[61]

Chaul has been identified as an early Portuguese shipbuilding site by some writers. But the information is not available in the Portuguese documentation.

Portuguese navigators used a variety of equipments for voyage from the Atlantic through Indian Ocean or Arabian sea to reach the western coast of India and on return. Different types of sails and oars were used to harness the natural and human energy for the propulsion of the vessels. In the absence of modern navigational equipments, they had their own equipments to decipher the direction based on the position of stars. They used astronomical techniques for navigation. They used special instruments and technique for the calculation of time, and depth of the sea. Messages from the captain in chief of a fleet had to be passed on to the various captains of the ships constituting the fleet. They employed some techniques for that. These aspects shall be discussed in the next chapter.

Thus it may be concluded that the shipbuilding activities that were quite advanced in the region of Calicut especially Beypore dwindled during the Portuguese domination on the Malabar coast. The constant fights between the Zamorin and the Portuguese paved the way for the decline of this industry in Calicut. But another centre, namely Cochin developed as a major shipbuilding centre during this period. This was chiefly because of the emergence of Cochin after the unprecedented flood in the Periyar river in the fourteenth century and the conscious effort made by the Portuguese to develop a major shipbuilding centre with the patronage of the king of Cochin. The port of Cochin was easily accessible through the network of lagoons and rivers to the hinterland from where better variety of timber could be brought for shipbuilding activities. Teak made available to the port of Cochin through these ways was another important ingredient in the development of the shipbuilding centre at Cochin. The geophysical conditions that developed under the Europeans later paved the way for the emergence of the natural port of Cochin and finally for the establishment of a modern shipyard to the detriment of the traditional centre at Calicut. It would however be difficult to deny the physical advantage of the port of Cochin vis-à-vis that of Calicut.

This short discussion brings home the following points:

a. the temperature of the water in the Indian Ocean demands a special type of timber to avoid woodworms.
b. the size of the ships and the tonnage are determined by the environment –arms and ammunitions, increased cargo and personnel.
c. use of iron nails to make the ship strong and durable.
d. a technology in keeping with the environment.

NOTES AND REFERENCES

1. This is not to forget the fragmentary contribution of the Italian shipwrights whose narratives are contained in the *Fabrica de Galere*, the original of which is lost. Two copies of it are available, one in the Bibliotheca Nazionale Centrale di Firenze, col. Mag., cl.xix, 7, *Fabrica de Galere* – probably of the fifteenth century – and another in the Australian National Library, Marco Foscarini collection Cod. 6391, *Arte di far galee e navi* (perhaps from the sixteenth century). The master shipwrights of northern Italy recited aloud the essentials of their projects to their pupils and assistants while the work progressed. Such verses were in course of time rendered to writing and collected to form the earliest written shipbuilding instructions, lyrical in form and based primarily on word and number. The authorities in northern Italy began to regulate 'state' shipbuilding by the thirteenth and fourteenth centuries through *decretti*/decrees related primarily to *Galley* building. The shipwrights developed specific directions from these documents. The oldest surviving shipbuilding manuscripts are based on these decrees dating from the early fifteenth century. Georgio Trombetta provides drawings of individual ship components showing the deliberate use of parallel perspective projections, which date from 1441 to 1449. These sketches were very small and drawn free hand: see Cottonian Mss.Titus, A.26, *Georgio Trombetta da Modon*, British Library, London, also Roger Anderson, 'Italian Naval Architecture about 1445', *The Mariner's Mirror*, II (1925), pp. 5-163. Pre Theodoro, a Venetian master shipwright, prepared a manuscript containing the forms of the frames and true-scale side elevation in the mid-sixteenth century. Two copies of his work have survived; one in the Biblioteca Nazionale Marciana, Ms. Ital. cl.iv, 26, Pre Theodoro de Nicolo, 'Instructione sul modo di fabricare galere', and another in the Archivio di Stato di Venezia, *Archivio Proprio Catarini*, 19, Theodor de Nicola, *Arte di far vaselli*. Joseph Furttenbach, a Genoa-based international merchant who operated between 1609 and 1617, described from his own experience the method of shipbuilding. His work was published in 1629 from Ulm (West Germany) under the title *Architectura Navalis*. Similarly Steffano de Zuane, a Venetian master shipwright, prepared another work on naval architecture. Add. Mss. 38655 Steffano de Zuanne de Michel *L'Architectura Navele* at the British Library, London. For further details on Italian *Galley* building, see Robert Gardiner, ed., *Conway's History of the Ship: The Age of Galley: Mediterranean Oared Vessels Since Pre-classical Times,* London, 1995, pp. 142-62. The English had some writing on shipbuilding in the sixteenth century. William Bourne wrote his *Treasure for Travellers* (1578); Mathew Baker wrote the incomplete *Fragments of Ancient English Shipwrightry* (Pepys Library Ms. 2820) in 1570. There are two more anonymous works in English on the subject, namely,

'A Treatise on Shipbuilding' and 'Treatise on Rigging' which remain unpublished. In 1626 the first English book on shipbuilding appeared under the title, *Accidence or the Pathway to Experience: Necessary for all Young Seamen* by John Smith. The same author published *Sea Grammar* in 1627. The French Jesuit Paul Hoste published *L'Arte des Armees Navales* in 1698. This contains a mathematical and geographical study on ships. The Dutch began publishing books on shipbuilding much later.

2. Some of these writings has been brought out with facsimiles in the wake of new publications related to recent celebrations of the Portuguese discoveries. They are: Fernando Oliveira, *O Livro da Fabrica das Naus*, Lisboa, Academia de Marinha, 1991. Manuel Fernandez, *Livro de Traças de Carpintaria*, Lisboa, 1989, João Baptista Lavanha, *Livro Primeiro da Architectura Naval*, Lisboa, 1996. There are a few additional documents of great value dated to the sixteenth and seventeenth centuries on the subject of naval architecture. Some were transcribed and published long ago as appendices and others still remain to be published. These comprise: Fernando Oliveira, *Ars Nautica* written in Latin in 1570, remaining in the manuscript form in Codex Vossiani Latini F. 41, ff. 1-236 in the Library of Leiden University; Marcos de Aguilar, mss. in the Bibliotheca da casa Cadavel written in 1640 under the title *Advertencias de Naveguantes,* photocopy of which is preserved in the Bibliotheca Central de Marinha in Lisbon; *O Tratado do que deve saber hu bom soldado para ser bom capitam de Mar e Guerra* by an anonymous author written in the second-half of the seventeenth century and located in the Mss. Biblitheca da Universidade de Coimbra. This material was published by Rocha Cadail in *Arq. Hist. Marinha I* (1936); *Coriosidade de Gonçallo de Sousa*, written by the end of the sixteenth century or the beginning of the seventeenth century. *O Livro Nautico ou meio pratico da construção dos Navios e Gales antigas e memorias de varias cousas* (end of the seventeenth century), Mss. Bibliotheca Nacional de Lisboa and published by Henriquez Lopez de Mendonça as an appendix in his *Estudos sobre Navios Portugueses nos seculos XV e XVI*, Lisboa, 1971; *Fragments of Ancient Shipwrightry* (c. 1586 Pepsian Library of the University of Cambridge).
3. Fernando Oliveira, *O Livro da Fabrica das Naus*, Lisboa, 1991, p. 134.
4. João Bapitsta Lavanha, *O Livro Primeiro da Architectura Naval*, Lisboa, 1996, p. 140
5. B. Arunachalam, 'Timber Traditions in Indian Boat Technology', in K.S. Mathew, ed., *Shipbuilding and Navigation in the Indian Ocean Region* AD *1400-1800*, Delhi, 1997, pp. 12-19. He makes mention of a Tamil palm leaf manuscript in a mutilated form under the title *Navoy Sastram*, which could be older than the *Kappal Sastram* of Tarangambadi (1620), that goes into details of timber quality: its physical defects, colour, appearance of a fresh-cut section of the log and so on. *Navoy Sastram* forms part of the McKenzie

collection in the Madras Archives. Arunachalam further refers to a Muslim Tamil *Cala-Vettu Pattu* of Nagapattinam that classifies timber as masculine, feminine and eunuch. *Kappal Sastram,* in fact, speaks of various varieties of timber that are useful for different parts of the vessel. *Karimarudu* according to Arunachalam was the most commonly used timber for the keel of a ship. This conclusion is arrived at through the extensive field study conducted by him and also based on the *Kulathurayyan kappal pattu.* He identifies several types of timber such as *punnai, aini, benteak,* jack tree, teak, *kongu, karunelli, vembu* and *karimarudu* used for various parts of a ship. He brings out the point that an ocean-going vessel was built not with one type of timber alone.

6. Oliveira, *O Livro da Fabrica das Naus, op. cit.*, p. 140.
7. Lavanha, *op. cit.*, p. 141.
8. *Ibid.*
9. Archivo Nacional da Torre do Tombo hereafter as ANTT (Lisboa), *Corpo Chronologico*, part I, maço 87, document 74; ANTT Gavetas, 15, maço 12, document 6; A.da Silva Rego, ed., *As Gavetas da Torre do Tombo*, vol. IV, Lisboa, 1964, pp. 384-5. This is a common sight in the Kallai and Chaliyam rivers.
10. *As Gavetas da Torre do Tombo,* vol. IV, p. 387.
11. Nycolão Gomçalves, *Livro em que trata das cousas da India e do Japão*, ed. Adelino da Almeida Calado, Coimbra, 1957, pp. 43-8.
12. 'Lembranças de cousas da India em 1525', in Rodrigo José de Lima Felner, ed. *Subsidios para a Historia da India Portuguesa,* Lisboa, 1868, pp. 21-31. Here the author gives the number of various types of vessels, such as *Galeões, Gales, Galeotas, Bargamtyns,* round ships with lateen sails, canoes, and small boats found in India around 1525. Some of them were built in Cochin.
13. Raymundo António de Bulhão Pato, ed., *Cartas de Afonso de Albuquerque*, tomo II, Lisboa, 1898, p. 137.
14. Oliveira, *op. cit.*, p. 144.
15. *Ibid.*, p. 146.
16. A local ritual song of the Malabar coast describes the way in which a carpenter selected suitable timber from trees of different sorts for various parts of a vessel like *pandi* (keel), *kal* (rib), *aniyam* (stern), *tattu* (floor), *kombu* (mast), *tandu* (oar), *chukkan* (rudder), *namkuram* (anchor) and *kalli* (compartment) built on the Malabar coast. C.M.S. Chandera, ed., *Kannakiyum Cheerammakkavum,* Kottayam, 1973 (Malayalam) and also K.K.N. Kurup, 'Indigenous Navigation and Shipbuilding on the Malabar Coast', in K.S. Mathew, ed., *Shipbuilding and Navigation in the Indian Ocean Regions* AD *1400-1800,* Delhi, 1997, pp. 20-5.
17. Archivo Nacional da Torre do Tombo (Lisboa), *Gaveta* 15, Maço 12,

document 6; Sousa Viterbo, *Trabalhos nauticos dos Portugueses, seculos xvi e xvii*, Lisboa, 1898, pp. 24-37.

18. Oliveira, *op. cit.*, pp. 146-8; Lavanha, *op. cit.*, p. 143.
19. *Ibid.*, p. 147.
20. *Ibid.*, p. 149.
21. Lavanha, *op. cit.*, pp. 145-6.
22. Oliveira, *op. cit.*, p. 149.
23. *Ibid.*, p.150.
24. Lavanha, *op. cit.*, pp. 148-67; Oliveira, *op. cit.*, pp. 146-211; Fernandez, *op. cit.*, pp. 117-212.
25. But it is curious to note that Ludovico di Varthema who spent some time in Calicut speaks of the way in which native ships were built with a lot of iron nails.ref. Ludovico di Varthema, *The Itinerary of Ludovico di Varthema of Bologna from 1502 to 1508*, London, 1928, pp. 62-3.
26. João de Barros, *Da Asia*, Decada I, Lisbo, 1778, pp. 424-32, Gaspar Correa, *Lendas da India*, tomo I, part I, Coimbra, 1922, p. 202; Fernão Lopes de Castanheda, *Historia do Descobrimento e Conquista da India pelos Portugueses*, livro I, Coimbra, 1924, pp. 83-5.
27. K.S. Mathew, ed.. *Shipbuilding and Navigation in the Indian Ocean Region AD 1400-1800*, Delhi, 1997, p. 40.
28. Lavanha, *op. cit.*, pp. 145-6; Oliveira, *op. cit.*, p. 151.
29. Lavanha, *op. cit.*, pp. 146-7; Oliveira, *op. cit.*, p. 192.
30. Letter of João Anes dated 6 January 1533 , in Viterbo, *op. cit.*, pp. 412, 419.
31. One *cruzado* at that time could be equal to 390 *reis* while 1 *pardao* fetched 320 reis.
32. Calado, *op. cit.*, pp.60-2.
33. Oliveira, *op. cit.*, p. 151.
34. One ton = *2 pipas* or 1.606 cubic metre. *Pipa* was a Portuguese tridimensional measure (Lavanha, *op. cit.*, p. 230). Eugenio Estanislaus de Barros, *Traçado e Construção das Naus Portuguesas dos séculos XVI e XVII*, Lisboa, 1933, pp. 12-13.
35. Oliveria, *op. cit.*, p. 164.
36. Constantino Barcellos, 'Construçaõ de Naus em Lisboa e Goa', *Boletim da Sociedade de Geografia de Lisboa*, 17ª serie, 1898-9, no. 1, pp. 29ff.
37. Biblioteca Nacional de Lisboa, 'Consultas resolvidas', Conselho Ultramarino no. 348, Numero de ordem 796, maço. 2; Barcellos, *op. cit.*, pp. 24 ff.
38. Raimundo Antonio de Bulhão Pato, ed., *Cartas de Affonso de Albuquerque* (hereafter *Cartas*), tomo II, Lisboa, 1898, p. 319.
39. A. da Silva Rego, ed., *Documentação para a Historia das Missões do Padroado Português do Oriente*, vol. 1, Lisboa, 1947, p. 141.
40. *Cartas,* tomo II, pp. 430 ff.
41. Correa, *Lendas da India*, tomo II, p. 139.

42. *Ibid.*, p. 488: Henriques Lopes de Mendonça, 'Estudos sobre Navios Portugueses nos seculos XV e XVI', in *Centenario do Descobrimento da America: Memorias da Commissão Portuguesa*, Lisboa, 1892, p. 10. This ship afterwards was sent to Portugal for the use of Dona Beatriz, the daughter of Dom Manuel I.
43. *Cartas*, tomo 1, Lisboa, 1884, p. 68.
44. Archivo Nacional da Torre do Tombo, Lisboa, Mss. *Nucleo Antigo*, no. 705, fl.13.
45. *Cartas*, tomo 1, pp. 303, 295.
46. A. da Silva Rego, ed., *As Gavetas da Torre do Tombo*, vol. X, Lisboa, 1974, p. 661.
47. Elaine Sanceau, ed., *Colecção de São Lourenço*, vol. II, Lisboa, 1975, p. 4.
48. Simão Botelho, 'Tombo do Estado da India', in Rodrigo José de Lima Felner, *Subsidios para Historia da India Portuguesa*, Lisboa, 1868, pp. 20-1.
49. K.S. Mathew and Afzal Ahmed, eds., *Emergence of Cochin in the Pre-Industrial Era: A Study of Portuguese Cochin*, Pondicherry, 1990, pp. 47, 58-9.
50. K.S. Mathew, 'Cochin and the Portuguese Trade with India during the 16th Century', *Indica*, March 1989, p. 80.
51. Historical Archives of Goa, Mss. *Livros das Monções*, no. 7 (1600-1603), fols. 113-15.
52. Trelado de huu Alvara do capitam moor Diogo Lopes de Sequeira. 'Per este ey por bem que Joanne Anes, mestre dos caprinteiros da ribeira de Cochym, aja de mamtymento em cada huu mees outro tanto como ha joão Luys , comdestabre dos bombardeiros, que he o que ele mesmo soya daver em tempo da° dalbquerque que Deus aja, e em tempo de Lopo Soarez segumdo tynha per seus alvaras, que lheu eu ora confirmey. E per este mando ao feytor desta feytoria de Cochim que lhe pague como aquy faaz memçam sem nenhua duvida nem embargo que lhe seja posto. Feito em Cochin aos xiiij de Junho de 1519.'
53. *Cartas* I, p. 295.
54. *Cartas* I, p. 253, letter written by Affonso de Albuquerque at Cannanore on 24 December 1513.
55. *Cartas*, I, p. 375.
56. *Cartas*, tomo I, p. 303.
57. *Ibid.*
58. 'Foime ditto que será meu serviço fazeremse nessas partes allguas náos pera servyrem nesta viagem da Imdia , asi pela espiriemcia que se tem das que se lá fazem durarem muito mais tempo que as que se fazem neste Reyno, como taobem por serem menos custosas e mais fortes, e irem faltando as madeiras pera as ditas náos; emcommendouos que vos imformeis se averá pesoas nese estado que queirão fazer alguas per comtrato asi nas partes do norte como em Cochim, qeu sejão de quinhemtas té seicemtas toneladas, e o preço porque darão acabadas cada hua das ditas náos, e que poderá

custar posta á vela; de que me avisareis', J.H. Da Cunha Rivara, *Archivo Portuguez Oriental Fasciculo* III, Delhi, 1992, p. 46.

59. 'A falta que ha de náos no Reyno pera a carreira da India he muito grande como deueis ter sabido por se terem perdido muitas, e irem faltando as madeiras per ellas; e porque sou informado que nessas partes se podem fazer muitas náos que são melhores e maes convenientes pera esta carreira que as que se fazem no Reyno encomendoues que procureis (como já vos tenho escrito outras vezes) por haver aluguas náos que estejão feitas de particulares, nouas, e boas, que possão servir nesta viagem, e ordeneis que se vão fazendo em todas , e a paga(?) dellas consignareis em alguas rendas minhas dessas partes não tendo dinheiro prompto pera se pagarem, e de qua se vos ajudará com algum depois que me avisardes das que fordes comprando contractando, e do custo dellas. E por El Rey de Cochin ter em seu Reyno muita copia de maderiras e officiaes, e se entender que lhe custarão menos as ditas náos a fazer que outra nenhuma pessoa, encomendouos que trateis com elle que dê toda ajuda e fauor pera estas náos, se fazerem, e se com elle mesmo quiserdes contractar que as dê feitas, seja com todas as seguranças necessarias....' Ref. J.H. Da Cunha Rivara, *Archivo Portuguez Oriental Fasciculo* III, Delhi, 1992, pp. 448-9.

60. François Pyrard Laval, *The Voyage of François Pyrard of Laval to the East Indies, the Maldives the Moluccas and Brazil,* vol. II, part I, London, 1888, p. 182.

61. For details of shipbuilding in Bassein, B. Arunachalam, *Mumbai by the Sea*, Mumbai, 2004, pp. 82-4.

CHAPTER V

Portuguese Claims to the Exclusive Domination of the Indian Ocean Regions

Portuguese navigators used a wide range of equipments for the voyage from the Atlantic through the Indian Ocean or Arabian sea to the west coast of India. Different types of sails and oars were used to harness the natural and human energy for the propulsion of the vessels. In the absence of modern navigational equipment, they had their own instruments to decipher the direction based on the position of stars. They had at their disposal astronomical techniques for navigation. They used special instruments and technique for the calculation of time and the depth of the sea. Messages from the captain in chief of a fleet had to be passed on to the various captains of the ships constituting the fleet. Ballast also had to be arranged.

As mentioned by Ibn Khaldun, medieval navigation depended on the wind. The natural energy was harnessed for propulsion. Sails of different types were employed to make use of the energy from wind.

Astronomical Navigation

Any art of navigation in which the mariners made observations of the positions of celestial entities like the sun, moon, stars, constellations of stars and planets for directing the movement of the vessels in the sea can be called astronomical navigation. Mariners through the ages were guided by the position of certain stars or constellations. But in the strict sense navigation began to be called astronomical if it was based on practical observation with a view to obtaining a coordinated horizontal position of the sun or of some star which enabled the pilot to choose with less uncertainty a convenient route after having had an approximate idea of the position of his vessel.

The information on the practice of navigation in early medieval Europe is very scanty. We may deduce from the words of the Roman poet Lucan (*c.* AD 65) that the Pole Star was favoured as an aid to navigation. We can reasonably be sure that by the end of the thirteenth century the seamen in the Mediterranean could have used a chart, a magnetic compass and sailing directions, possibly also a sand-glass. Directions were given in the sailing directions, according to the traditional wind directions, as winds, half- or quarter-winds, and the distance sailed was measured merely by estimating the ship's speed and measuring it. In the thirteenth century rudimentary trigonometric tables were applied to navigation. It seems that the Mediterranean sailor was fortunate in comparison with his more northerly contemporaries who had to make do with a floating magnetized needle and the deep sea lead.

The first reference to astronomical navigation, as explained above, in Portugal dates back to the third quarter of the fifteenth century.[1] Navigation of this sort was based on observation of the Pole Star and the sun, made with the quadrant and the planispheric astrolabe in general use among the medieval astrologers.

Ibn Khaldun (b. Tunis 1332: d. Cairo 1406) wrote in 1377 in the prolegomena of his *Muqaddima* about the problem of navigation:

> Navigation on the sea depends on the winds. It depends on knowledge of the directions the winds blow from and where they lead, and on following a straight course from the places that lie along the path of a particular wind. . . . The countries situated on the two shores of the Mediterranean are noted on the chart . . . which indicates the true facts regarding them and gives their positions along the coast in the proper order. The various winds and their paths are likewise put down on the chart. . . . It is on this . . . that (sailors) rely on their voyages. Nothing of the sort exists for the Surrounding Sea. Therefore, ships do not enter it, because were they to lose sight of shore, they would hardly be able to find their way back to it.[2]

This situation was ameliorated in due course of time. It was necessary to be able to determine, reasonably accurately, one's position on the open sea, in fact to find one's latitude and longitude. This meant adoption of astronomical techniques besides improving the magnetic compass and of the marine chart both available till then in rudimentary form. These changes in techniques began in the late fifteenth century.

An illustration in a manuscript of *c.* 1450 of Heinrich Suso's *Horologe de Sapience* [3] depicts a range of time-telling devices so as to symbolize the title of Suso's work. There is a large mechanical clock (weight-driven and controlled by a verge and foliot escapement), a small table clock (spring-driven), three sundials, a horary quadrant, a planispheric astrolabe, and a bell-ringing device. Except for the latter, all these instruments were ultimately to influence the progress of navigation. Below I describe a few instruments of navigtion used prior to the voyage of Vasco da Gama to India.

Armillary Sphere

This was an instrument used by D. Manuel. The Ptolemaic geocentric system was shown by a small globe -- the earth -- in the centre of the celestial sphere which is delineated by rings representing the polar circles, the tropics, the equator and ecliptic. It is of engraved brass with a globe of wood; it is probably of the fifteenth century. Deriving from Ptolemy's *astrolabon*, the armillary sphere in Islam appears mainly to have been, like Ptolemy's instrument, an observational instrument. To be effective it had to be large. In the European tradition, the armillary sphere is not primarily an observational instrument. It was primarily used in Europe for demonstrational or didactic purposes. An elaborate sphere with star pointers attached to the rings could show the apparent rotation of the stars about the pole and serve also for the solution of simple problems in spherical trigonometry. An instrument derived in the sixteenth century from the armillary sphere became a useful nautical instrument.

Equatorium

Equatorium was a geometrical calculating instrument used to determine the positions of the planets according to the Ptolemaic mathematical theory.

Torquetum

This was an observational instrument giving direct readings in equatorial or ecliptic coordinates. It was devised at the end of the thirteenth century.

Navigational Guidelines

The Portuguese after entering on overseas expansion developed navigational methods based on their accumulated knowledge and experience and issued guidelines (*Regimentos*) for navigation which in fact proved useful to other Europeans too. Some of them are mentioned here below:

Regimento of the Pole Star

The navigators used Pole Star to know the height of the pole or latitude in the Eastern Seas. It was measured with the fingers or *isba*. The Portuguese navigators prepared Guidelines (*Regimento*) for referring to the position of Pole Star which is reproduced in the Manual of Munich of 1483/4.[4]

Regimento for Calculating Night Hours by Pole Star and Guards

The 'little bear' (*Ursa menor*) was used to determine the night hours. When the 'Advance Guard' moved around the Pole, it looked like a unique needle of a huge clock and hence it was used to find the night hours, at least since the thirteenth century.

Regimento of the Height of the Pole by the Southern Cross

The Portuguese identified the astral constellation, the Southern Cross and worked out a *Regimento* of the height of the pole by it for navigation in the southern hemisphere. Pilot Cadamosto named it 'Cruzeiro' in 1455. Pilot Pero Anes who travelled to India in the fleet of Viceroy Francisco de Almeida in 1505 referred to it as 'southern cross'. Along with João de Lisboa, he carried out a joint experimentation by the end of 1507 and subsequently this *Regimento* was reproduced in all the later works.

Regimento of the Height of the Pole by Other Stars

The pilots on landing used to observe other stars which enabled them to fix the exact moment of the day in the maximum or minimum rising of the horizon.

Regimento of the Height of the Pole by the Sun

The pilots based on Sagres studied experimentally the possibility of finding out the ship's latitude in good weather by the observation of the sun. They obtained important results in navigation. The information about this is incorporated in this Regimento.

Regimento of the Astrolabe (1483)

The board of mathematicians appointed by King John II of Portugal prepared a *Regimento* of the astrolabe by showing their knowledge of cosmography, use of astrolabe and table of solar declination. It contained rules for determining latitude by the height of the sun and the rule of the Pole Star to measure the route covered. It also contained a list of latitudes, a calendar with the table of the position of the sun in the rising of the Zodiac, and a calendar of tides.

Roteiros (Logbooks)

Roteiro meant the day-to-day report of a voyage written by the navigator mentioning the route, distance covered, astronomical observations and calculations, navigational particulars, adventures during the voyage, disembarkation, description of the land visited and so on. The Portuguese *roteiros* were the forerunners of the hand-books for the pilots. There a couple of famous Roteiros which are mentioned below.

ROTEIRO OF DUARTE PACHECO PEREIRA (*ESMERALDO DE SITU ORBIS*)

Duarte Pacheco Pereira came to India in 1503 and worked in Cochin as the captain-in-chief. He fought numerous battles in defence of Cochin against the forces of the Zamorin. On his return to Lisbon he wrote a *roteiro* between 1505 and 1509 at the orders of the king. It was entitled *Esmeraldo de Situ Orbis*. It is a result of a laborious study and exhaustive survey on cosmography and geography. It has sixteen cartographic representations, landscapes, world maps, nautical astronomy and related studies.[5]

ROTEIRO OF JOÃO DE LISBOA (*LIVRO DE MARINHARIA* 1519)

João de Lisboa made several trips to India starting with the historic voyage with Vasco da Gama. His work has a precious collection of *roteiros* titled 'Book of Routes from Lisbon to India'.

ROTEIRO OF ANDRÉ PIRES (*c.* 1530)

The book of André Pires contains a treatise on the compass, rules about the Southern Cross and other valuable nautical information. It basically deals with the routes from Portugal to India.

ROTEIRO OF DIOGO AFONSO (*c.* 1536)

His work contains information about the working of magnetic needle in various regions as well as the landmarks and signs derived from the sight of birds and marine planets.

ROTEIROS OF D. JOÃO DE CASTRO

D. João de Castro (1500-48) wrote three *roteiros*, the first of which is called *Roteiro from Lisbon to Goa* (from 6 April 1538 to 11 September 1538), the second is the *First Roteiro of the Coast of India from Goa to Diu* (from 21 November 1538 to 29 March 1539) and the third, *The Roteiro from Goa to the Red Sea* (31 December 1540 to 21 August 1541). The profusely illustrated *roteiros* contains vast knowledge of nautical experience, observations of latitude, calculations of longitude, variations of the magnetic needle, tides, eclipses, winds and currents.

ROTEIRO OF MANUEL DE ALVARES (1545)

Manuel Álvares was the pilot of the ship *Grifo* under Governor João de Castro on the voyage from Lisbon to Goa in 1538. He completed his *roteiro* in 1545.

ROTEIRO OF PERO VAZ FARGOSA (1560)

Pero Vaz Fargosa wrote his *roteiro* of the India voyage in 1560.

ROTEIROS OF VICENTE RODRIGUES (1575 AND 1591)

Vicente Rodrigues spent a life time in the Carreira da India. He left Lisbon on 7 April 1568 as the pilot of the *Chagas* under Viceroy

Luis de Ataide. Again in 1590 he was the pilot of the *Bom Jesu* under Viceroy Matias de Albuquerque. His second *roteiro* was written in the course of this voyage.

There are a few more *roteiros* containing similar pieces of information.

The Portuguese navigational activities passed through three different stages, namely, *Marinharia* (Mariners' art), *arte de navegar* (art of navigation) and finally nautical science. There are a couple of books in all these phases of development: *Le Livro de Marinharia de Gaspar Moreira; Livro de Marinharia de Manuel Álvares, Arte de Navegar de Manuel Pimentel* and *O Livro de Mariharia de André Pires.*[6]

Instruments of Navigation (Astronomical)

It is reported by the Portuguese scholars that the earliest reference to astronomical navigation based on the Pole Star, and of the Sun taken with the quadrant and the planispheric astrolabe in the Atlantic dates back to the third quarter of the fifteenth century.[7] We shall discuss in the following pages details about these instruments used by the Portuguese for astronomical navigation.

PLANISPHERIC ASTROLABE

It is the best known and one of the oldest of medieval scientific instruments. It was used observationally for time telling by day or night, and in surveying. In modern terms, it is an analogue computer, serving to solve astronomical problems by simulating the apparent rotation of the stars about the pole. It may be considered an armillary sphere which, by means of stereographic projection, in this case from the South Pole on to the plane of the equator, results in angular measurements from the centre remaining undistorted.

THE MARINER'S QUADRANT

The earliest surviving European quadrant dates from 1300. It is similar to what is known as *quadrans vetus* dating back to the twelfth century. It was primarily used as a sundial. It had a plumb-line, with a sliding bead and weighted with a bob, hung from the apex. The first known representation of a quadrant meant for use by the mariners is found in the posthumous edition (1563) of the *Reportório dos tempos* (first published in Lisbon in 1518) by Valentine Fernandes. The illustration at the beginning of the chapter is entitled as '*o regimento pera se poder reger pelo Quandrante ou Astrolabio pela estrella do Norte.*'[8]

The quadrant illustrated by Velentine Fernandes has a horary scale (unequal hour diagram) which has been reduced to occupy a much smaller area of the instrument. The simple scale of degrees along the arc has been emphasized by lengthening the divisions at 5° and 10° intervals. This instrument was chiefly used to measure the altitude of celestial body, e.g. the Pole Star. On leaving Lisbon, the pilot was instructed to mark where the plumb-line fell while observing the Pole Star with the Guards of the Little Bear east-west in relation to that star. If later on he wanted to know the distance from Lisbon, he whould find the difference in degrees between the original position of the plumb-line and that of a new observation, and convert to distance on the basis of 1° = 16⅔ leagues.

THE MARINER'S OR SEA ASTROLABE

The mariner's or sea-astrolabe was reduced to a graduated circle of brass or wood with an alidade and a suspension ring while the nautical quadrant was simply a sector of quarter of a circle, graduated along the arc, with two sights on a radius, and a plumb-line.[9] The mariner's astrolabe is not a planispheric astrolabe.[10] There is no stereographic projection of the celestial sphere, nor a zodiac/calendar scale, a shadow-square, or a horary diagram on the back. It consists simply of an alidade moving over a scale of degrees for measuring altitudes. The body of the astrolabe is very thick and it is therefore heavy and the thickness increases towards the bottom, so that it hangs well. The body is perforated so as to offer a minimum resistance to the wind. The alidade has the sights set closely together to facilitate the taking of solar readings. The mariner's astrolabe was developed during the last decades of the fifteenth century.[11] Vasco da Gama for his first voyage to India had a wooden astrolabe and small brass astrolabes as described by João de Barros in *Decadas da Asia* measure the height of the Sun.[12]

The earliest illustration of an astrolabe for nautical use is found on a chart drawn by Diego Ribero in 1525.[13] This depicts an astrolabe made from a solid plate of metal or wood. There is a shadow-square in the lower half. The drawing could equally well show the back of a planispheric astrolabe. The link between a planispheric astrolabe and the mariner's astrolabe appears to extend no further than the identity of the names.[14]

THE NOCTURNAL

Nocturnal or *nocturlabio* was developed during the sixteenth century. It is an instrument for determining the time at night by observation of the apparent rotation of α or β (the great guards) *Ursae majoris* or of β *Ursae minoris*, about the Pole. Though there is scientific history of its origin, some would perhaps seek to derive it from the very ancient Chinese circumpolar constellation template.

Towards the end of the thirteenth century, Ramon Lull described an *astrolabium nocturnum* or *sphaera horarum noctis* consisting of a single disc, perforated at the centre and engraved with concentric scales of the months and the 24 hours, correlating the midnight position of β *Ursae majoris* throughout the year. The time could be found by counting the hours before or after the midnight position of the time of year. This is done by sighting the Pole Star through the central hole and noting against which hour the star stood. Robert Norman in the sixteenth century described an equally crude nocturnal which consisted of a small iron ring held firm at the centre of a larger ring by four equally spaced radial threads: the larger ring had 24 or 32 equally spaced spikes on its circumference and was aligned on the celestial meridian by either of a pair of threads which formed a diameter. The instrument which had to be used with a table of midnight positions throughout the year is a curiously late survival of the most elementary form of nocturnal, by then long superseded. The earliest known nocturnal as a separate instrument is the one made in 1511 by Laurentius Vulparia of Florence. Nocturnals are fairly common instruments in the sixteenth and seventeenth centuries. The prime purpose of a nocturnal was not time-telling, but finding the location of the Pole from an observation of the Pole Star according to some scholars.

THE AZIMUTH COMPASS

Azimuth meant an angle related to a distance around the earth's horizon, used to find out the position of a star or planet. Azimuth

Compass was an instrument which derives ultimately from an *instrumento de sombras* (shadow instrument)[15] described and illustrated in the *Tratado da sphera* of Pedro Nunez,[16] published in Lisbon in 1537. João de Lisboa in 1514 wrote on magnetic variation in his *Tratado da Agulha de marear.*[17] Felipe Guillen produced in 1519 an instrument for finding magnetic variation and Magellan was also provided with an azimuth-measuring instrument. Nunez' instrument was used to measure azimuths of the sun when at equal altitudes before and after midday. If any difference between the azimuths was found, this was halved to give the variation of the compass-needle from the meridian. The instrument was hung in cords or mounted in gimbals to insulate it from the movement of the ship. Gimbals are sometimes known as the Cardan suspension. Gimbals were known to medieval technology. It was in the sixteenth century that the foundations were laid for the science of terrestrial magnetism with its far-reaching effects on the design of the marine compass.

Mechanical Clock as an Aid to Navigation

The measurement of magnetic variation could not contribute to the solution for the problem of intractable longitude. Gemma Frisius published the work *De principiis astronomiae et cosmographiae* in 1530 at Louvain and Antwerp which suggested some means whereby a solution could be made available. Small clocks or watches from Germany (Nürnberg) were used for measuring longitude though they were not practical. They were inaccurate as time-keepers. It was against this background that Frisius suggested some modification and recommended the use of a small portable clock. Yet even this was not very practical.

The Universal Equinoctial Ring Dial

Gemma Frisius invented an astronomic ring which included facilities for latitude adjustment, sights on the movable ring, equatorial and meridional fixed rings and star names. This did not become very popular but it inspired the invention in the seventeenth century of the only form of sundial that was of much use to seamen, the universal equinoctial ring.

Substitutes for Quadrant and Astrolabe

The Portuguese navigators of the sixteenth and seventeenth centuries adopted a few instruments for measuring altitude as substitutes for the astrolabe and quadrant. This was done to avoid drawbacks found in the use of these instruments such as the cross – staff, the *kamal*, nautical ring (*anel náutico*), shadow instrument, nautical ring of another type (*armila nautica*) and the graduated semi-circle.

THE CROSS STAFF (*Balestilha*) AND JACOB'S STAFF

Cross staff was used for altitude-measuring for navigation and came into use in the sixteenth century. Even if this was already in existence and was probably invented by Judaeo-Provençal philosopher and scientist Levi ben Gerson (1288-1344), its use became popular in the sixteenth century only. The Jew described it in his treatise in 1342. There is no evidence to affirm that The Cross Staff was used for navigation before the sixteenth century. It was João de Lisboa and André Pires writing in 1520 who made reference to it for the first time for navigation.[18] The cross staff used in navigation differed slightly from the cross staff described by the Catalan Jew by name Levi ben Gerson. The one employed for navigation had a scale which gave readings in degrees, and it is the origin of this form which is of particular interest in the history of navigation. The earliest surviving nautical cross-staff is probably the wooden instrument left by Jacob van Heemskirck and Willen Barenstszoon in the 'Behouden Huis' at Nova Zembla in 1596 or 1597. It has been suggested that the navigator's interest in the cross-staff was aroused by acquaintance with the Islamic *kamal* (also known as the *tavoletas da India* mentioned both by João de Lisboa and André Pires).[19]

The use of the cross staff in navigation has not been hinted in the writings of Mestre João who was charged with the task of checking the precision of the various instruments meant for measuring altitude in the voyage of Pedro Álvares Cabral in 1500. No mention is made in the *Esmeraldo de Situ Orbis* by Duarte Pacheco Pereira.

Nor do we find any reference to it in the writings of Gaspar Correa while he reproduces the discussions made by a Jew named Çacuto in 1502 before Dom Manuel, king of Portugal, on navigation and navigational instruments.[20] Luís de Albuquerque states that the Portuguese navigators did not use cross staff before 1518 since this has not been mentioned even in the list of instruments supplied to Fernão de Magalhães (Magellan) for his voyage round the world. The list contains 21 quadrants and 7 astrolabes. João de Lisboa (*Livro de Marinharia*) for the first time refers to the use of the cross staff in navigation for making solar observations. At any rate it was in use on Portuguese ships before 1529. André Pires makes reference to *kamal* or *tavoletas da India* in the first half of the sixteenth century as the 'Moor's Cross-staff' (*balhistinha do mouro*).[21] Pedro Nunes had already referred to it by name in the *Tratado da Esfera* published in 1537.[22] By the second half of the sixteenth century the use of the cross-staff became common and it was preferred to the astrolabe.

Luís de Albuquerque distinguishes cross-staff from Jacob's staff, though the former was also in error called Jacob's staff.[23] There are references to the Jacob's staff before the end of the fifteenth century. S. Gunther found only one work dated before the second half of the fifteenth century that referred to Jacob's staff. Its author was the Franciscan Theodorico Ruffi who wrote the work *Baculus Geometricus alias baculus Jacobi*.[24] It gives the description regarding its use in finding distance and inaccessible heights. It is primarily an instrument for measuring distance while the cross staff measured the angle in degrees of arc with a precision which theoretically could be as high as one degree. It was chiefly meant for measuring altitude.

Jacob's staff according to the description given by Sebastian Münster (1551)[25] was made of two parts, a rod or staff six palms in length, which in Portugal and in Spain was called a *virote, verga, flecha* or *radio*, with a scale dividing it into six or eight parts; and a cross-piece called the cross, *soalha* or *sonaja* (Spanish), *transvesario, franja* or *mortinete,* equal in length to one of the parts into which the staff was divided, but wider, and with a hole in its centre. The staff fitted into this hole so that the cross could slide along it.[26]

The following procedure was employed to find the distance between two faraway points distant inaccessible to the observer:

The cross was adjusted to one of the divisions of the staff and the observer drew near to, or away from, the distance to be measured until he reached a point where, looking from one of the extremities of the staff, he could sight, at the extremities of the cross, the two points between which the distance was to be measured. This position was marked on the ground and the observer moved the cross to the next division along the staff and found a new position where he could obtain a similar sighting. The distance between the two positions gave the distance required. Five or seven angles in two directions or altitudes could be obtained with this type of Jacob's staff depending on whether the staff was divided into six or eight equal parts. Gemma Frisius distinguished the cross staff from Jacob staff. He called the latter *Radius Astronomicus*.[27]

The Cross Staff was used for some time in Portugal to measure the meridian altitude of the sun as we can deduce from the work of João de Lisboa where a 'regiment to take the sun with the cross staff' is given.[28]As sunlight would have blinded the observer if he tried to sight its centre with the naked eye, the books of instructions give some solutions.

To sight the sun in such a way as to cover it with the cross. In this case it was necessary to deduct from the observed altitude half of the apparent diameter of the sun, calculated at 15′.

Or take, the sight with the back towards the star by adding to the eye-end of the cross-staff a reflector with a slit through which the horizon could be seen. The sight was made with the eye placed at one end of the cross.

Pedro Nunes in his *De arte atque ratione navigandi* strongly condemns the use of the cross-staff to take altitudes. However, the sailors went on using the cross-staff on board the ship.[29]

KAMAL

It is held by Portuguese authorities on navigation that Vasco da Gama's sailors brought the *kamal* to Europe where it was tried out for some time. It was used under the name *tavoletas da India* or Moorish cross-staff.[30] The anonymous Florentine gentleman who accompanied Vasco da Gama and wrote an account of his voyage makes two references to a certain 'wooden quadrant'. It is generally accepted that these words referred to *kamal*.[31] Master João who was charged to make observations with the *kamal, astrolabe*, and the *quadrant* during Cabral's voyage in 1500, makes specific reference to the *tavoletas da India* in a letter addressed to Dom Manuel, king of Portugal. This correspondence indicates that no opportunity was lost to try out the *kamal* on the first long voyage after Vasco da Gama's return.

The *kamal* was made of one or more square or rectangular tablets through the centre of which was threaded a piece of string with knots at intervals from the tablet. One chose the knot on the string which, held close to the eye of the observer or between the teeth placed the tablet, when, held in front of the eye with the string taut, at such a distance that the pilot could sight the horizon along the lower edge of the tablet and the star at the upper edge to obtain the altitude of a star with this instrument.

THE BACKSTAFF

The problems encountered in the use of the cross-staff were surmounted by the use of the backstaff. Thomas Harriot (1560-1621), the English mathematician and friend of Sir Walter Raleigh showed how to correct the parallax error which arises because the eye is not on the axis of the staff. Harriot sought to redesign the instrument, but it is with the name of John Davis, *c.* 1595 that the improved version, called a backstaff (or Davis Quadrant) was associated. The method of use accounts for the name. The backstaff ultimately inspired the invention of the sextant, by way of Robert Hooke's reflecting instrument of 1666, Isaac Newton's instrument of 1700 and the octants of John Hadley (1682-1744) and of Thomas Godfrey (1704-44) of Philadelphia.

THE SECTOR

The increasing accuracy in navigation in the sixteenth century and the consequent mathematization of the techniques involved resulted in the use of a number of ancillary instruments. One of these was the sector which was gradually evolved during this century from the proportional compass. It is based on the principle of equal triangles. The sector is usually associated with the name of Galileo. It carried information on the scale for gunners.

NAUTICAL RING (*ANEL NÁUTICO*)

Nautical ring, known also as the astronomical ring or graduated ring is first mentioned in one of Pedro Nunes' works in the chapter on instruments with which altitudes of the stars can be obtained.[32] It was another instrument with which the altitude of the sun could be found. The credit of inventing it goes to Pedro Nunes the cosmographer of Portugal appointed in 1529. The instrument which was used for solar observation, consisted of a ring, rectangular in cross-section, fitted with a suspension ring at a particular point. The ring should be 'a third' in diameter and the thickness of 'a finger' according to Garcia de Cespedes while Manuel Pimentel furnishes the dimensions as three quarters of a span (*palmo*) in diameter and one inch (*polegada*) in thickness.[33] The observer had to suspend the nautical ring by means of a ring and rotate it until its plane was directed towards the sun. Once in position, sunlight passing through the hole fell on the point of the graduated arc, where the required coordinate could be read. The degrees of altitude are, on this instrument, twice as large as they would be if there was a centrally pivoted alidade, as we find on astrolabes of the usual type. The ring invented by Pedro Nunes did not find continued or lasting use in navigation.

ARMILLA NAUTICA

There was mention of another nautical ring or *Armilla Nautica* referred by Father Francisco da Costa who used to teach in the Jesuit Colégio de Santo Antão in Lisbon. This was used to take the altitude of the sun by mariners at sea. It consisted of a disc similar to the *mater* of an astrolabe, and like an astrolabe, was fitted with a suspension ring. The instrument gave either coordinate directly; all that had to be done was to hand the *armila* vertically from its suspension ring, and to direct the edge of the disc bearing the style towards the sun. The shadow cast by the style upon the graduated scale of the disc gave a direct reading of the required coordinate.

THE GRADUATED SEMI CIRCLE

Manuel Pimentel was the first to describe this instrument.[34] As the name indicates, it consisted of a semicircle of metal. This was used to find out the altitude of stars but not frequently.

Others

During the second half of the seventeenth century, other altitude instruments began to appear in Portuguese navigation, but the greater part of those known were imported from England without any contribution from the Portuguese instrument-makers. The double-arc quadrant or Davis quadrant developed by John Davis at the end of the sixteenth century is one among them. Semi-quadrant and octant were used in navigation. Another instrument known as *Nonius* was also in use for navigation.[35]

(A) SHADOW INSTRUMENT

Among the instruments recommended by Pedro Nunes for observation of altitudes of the sun there is one called the 'instrument laying on the plane' which was employed by D. João de Castro. This was properly referred to as the 'shadow instrument'. It consisted of a flat rectangular board (base) This was used to know the altitude of the sun. There are frequent references to the practical use of the instrument in the *Roteiro de Lisboa a Goa* by D. João de Castro. The instrument he used was given to him by Prince Luis, who appears to have been a fellow student. Two shadow instruments were given to him. One was used to determine the declination of the magnetic needle (for which the sun's altitude had to be known): the other was used by Castro to determine the altitude of the sun in order to find if the rule given to us by Doctor Pedro Nunes were true and certain enough for us to know the altitude of the Pole at every hour or the day when a shadow was cast.

(B) MAGNETIC COMPASS AND MAGNETIC DECLINATION

João de Lisboa in his *Tratado da Agulha de marear* (1514) provides details on the magnetic compass. It consisted of a compass-rose, with 32 points, cut out of thick paper or pasteboard; a magnetic needle to be mounted over or under the compass-rose; and a box into which these two fitted and which, some years later, was to be suitably suspended in a second or outer box. The needle comprised two iron or steel blades (called *ferros*) of equal size and pointed, but slightly curved at the centre, so that only the points touched. The two 'irons' were not permanent magnets. So they had to be magnetized from time to time by being rubbed with a lodestone, as mentioned by João de Lisboa. The needle should be large in size and as uniform as possible in weight and shape. The needle was firmly fixed to the compass-rose. North was marked on the latter by *fleur-de-lys*; and often a cross drawn at the end of the line pointing east showed the direction of Jerusalem in relation to Western Europe. Both rose and needle were placed in a box so that their centre rested on a turned pivot fitted at the centre. The cylindrical inner wall of the box was divided into 32 parts corresponding to the rhombus of the compass-rose. Later this division was made in degrees. The compass as described by João de Lisboa had to be placed on a platform slung from ropes near one of the ship's masts, a practice which Pedro Nunes still refers to. Left to the force of gravity, the platform remained more or less horizontal and the needle practically level.

Some of the Portuguese nautical charts of the beginning of the sixteenth century include a scale of latitudes drawn from north to south, in continuous line divided into two parts by the equator.

Seasons of Navigation

Departure from Portugal

The first fleet made ready for discovery of the direct sea-route to India under Vasco da Gama left Lisbon for India on 8 July 1497 with four vessels, namely *São Gabriel* commanded by Vasco da Gama, *São Rafael* under Paulo da Gama, brother of Vasco da Gama, *Bérrio* lead by Nicolau Coelho. It is believed that the fleet anchored off Kappad on 18 May 1498.

The fleet of Franscico de Almeida left Belém, Lisbon on 25 March 1505.[36]

The ship *Conceição* left Lisbon for India on 19 May 1516 and reached Madeira Island and from there on 22 May proceeded to India and reached Goa on 16 August.[37]

The ship *São Leão* left Lisbon on 10 April 1522 and reached Bhatkal on 15 October 1522. It was reported that the ship *São Leão* reached Gran Canária on 1 March and reached Mozambique on 18 August. She left Mozambique for India on 25 August.[38]

The ship called *São Roque* left Lisbon on 2 April 1529 and reached Mozambique on 15 July and Goa on 25 August 1529.[39]

Another ship *A Graça* by name left Lisbon on 10 April 1532.[40]

Departure from India

The fleet left Pantalayani on 29 August 1498 for Portugal.

A ship named *Santo António* left Cochin on 5 February 1523 for Portugal. On the way she reached Mozambique on 8 October and left for Lisbon on 22 October and finally entered Lisbon on 30 February 1524.[41]

Exclusive Claim

The Portuguese claimed exclusive rights for the navigation of the Indian Ocean region through several arguments. We shall have a glance at their claims.

> Mare igitur proprium omnino alicujus fieri non potest, quia natura commune hoc esse no permittit, sed jubet, immo ne litus quidem;. . . . Omnes igitur vident eum qui alterum navigare prohibeat nullo jure defendi. . . .[42]

A Dutch lawyer Hugo Grotius by name argued in 1604-5 that the Indian Ocean and the navigation thereof could not be appropriated by anyone. In support of his views he brought out the opinions of jurists and philosophers as cited above. He vehemently challenged the claims laid by the Portuguese for the exclusive possession of the Indian Ocean and the navigation therein. In fact the Portuguese tried to establish their exclusive right over the Indian Ocean and the maritime trade in the Indian Ocean regions by taking practical steps and also propounding their arguments based on various titles. This study is aimed at examining them in the light of the contemporary sources.

Interested in maintaining a monopoly in the Indian Ocean regions, the Portuguese asserted their exclusive right of passage. The king of Portugal at the beginning of the sixteenth century assumed a controversial title, 'Senhor da Navegação e conquista da India.' Others interested in navigating in the Indian Ocean were constrained to take a pass from the Portuguese. This was extended even to Indians. So, some discussions on this aspect shall be found in this chapter.

Papal Authority

Right from the beginning of the activities of expansion and maritime explorations, the Portuguese prepared themselves for the monopoly. As religion and religious heads held great importance in the late medieval Europe, they saw to it that with the intervention of the Popes rivals were kept away from the Indian ocean regions and the territories they were planning to discover and conquer. The Portuguese, intent on obtaining exclusive right over the Indian Ocean regions and proprietary claim over the lands and seas to be discovered in course of time, equipped themselves with papal bulls which permitted them to discover, conquer, and appropriate whatever area they liked. Whenever they were challenged by anyone in connection with their claim, they referred to the papal authorization and argued for their special right. Therefore, it will be worth examining the tenor of papal bulls adduced by the Portuguese in this regard.

The organized plan for the overseas expansion of the Portuguese was executed by the military order of Christ which received ecclesiastical approval from Pope John XXII on 15 March 1319.[43] The Portuguese Kings Edward and Affonso granted the jurisdiction over the conquered territories to the military order of Christ and its grand master, Infant Dom Henrique which was confirmed by Pope Eugene IV through his bull *Etsi Suscepti* issued on 9 January 1442. Subsequently, King Affonso V of Portugal obtained authorization from Pope Nicholas to conquer lands occupied by non-Christians anywhere in the world and bring them to subjection. All their possessions in terms of movable and immovable goods could be appropriated by the Portuguese. They were also permitted to enslave the enemies of Christ perpetually in the light of the bull *Dum Diversas* issued by Pope Nicholas V on 18 June 1452.[44]

A document of great significance on the basis of which the Portuguese argued for exclusive rights for overseas possessions especially the Indian Ocean regions was issued by Pope Nicholas V on

8 January 1454. The papal bull entitled *Romanus Pontifex* is rightly interpreted as the charter of Portuguese imperialism by some historians. The Pope appreciated the overseas activities of Dom Henrique, the uncle of Affonso V King of Portugal, since 1419 which promoted, to a large extent, the salvation of souls and glory of God, the Creator. He shared the view of Henry the Navigator that the ocean up to India was not crossed by anybody and that Indians worshipped the name of Christ was held in great esteem by the Pope.

The Pope appreciated the opinion of the prince that the Portuguese could circumnavigate the African continent and with the help of Indians who were reported to be believers in Christ, they could bring to subjection not only the Muslims but also all the others who did not believe in Christ and were not affected by the religious tenets of Muslims. The possibility of preaching in the name of Christ to non-Christians because of the extraordinary pains taken by the Portuguese was highly commended by the Pope. Therefore he permitted them to invade, conquer and appropriate territories and kingdoms of all those who were outside the Christian faith and reduce them to perpetual slavery. Full power in this regard was granted to Dom Henrique, Affonso, the King and all his successors. This was done by the Pope *motu proprio* within the fullness of his apostolic power and wisdom as he himself made it clear in this bull. The Pope further approved the Portuguese monopoly of navigation and trade, and extended it to India. Others were forbidden to interfere in any way with the Portuguese conquest and trade under penalty of excommunication.[45]

The Portuguese, keen on making known to others the full purport of the Papal largesse, organized a public proclamation of the bull in the Se Cathedral, Lisbon on 5 October 1455 before a large congregation. The French, the English, Castilians, Galicians and Basques were invited to witness the ceremony wherein both the Latin original and the Portuguese translation were made available to the public. This was a calculated move to keep others off the areas of Portuguese interest.

The monopoly over the Indian Ocean regions claimed by the Portuguese was further buttressed by Pope Calixtus III through his bull *Inter caetera* issued on 13 March 1455.[46] Confirming the order of Pope Nicholas V, the new Pontiff conceded to the order of Christ under Prince Henry, the spiritual jurisdiction over the areas conquered and to be conquered by the Portuguese. The Grand Prior of the order was given the power to nominate the incumbents to all the benefices both spiritual and temporal. Thus the Order of Christ was given a very extensive jurisdiction overseas. On the death of Prince Henry in 1460, the King Affonso was made the Grand Master of the Order of Christ through the bull *Dum tuam* issued by Pope Pius II on 26 January 1460.[47] Pope Sixtus IV confirmed the bulls of Nicolaus V and Calixtus III on 21 June 1481 through his order popularly known as *Aeterni Regis.*[48]

They saw to it that their neighbouring rulers of Spain too kept off the territories that would be discovered by them. The Spanish monarchy that became stronger after the marriage of Ferdinand of Aragon and Isabella of Castile in 1469 and especially after the *reconquista* and annexation of Granada in 1492 entered the scene of explorations. The maritime activities of discovery sponsored by the Catholic kings of Spain to find a safe sea-route to India without trespassing on the Portuguese preserves were conducted by sailing northward across the Atlantic. No sooner did Columbus under Spanish patronage make the landfall in the Caribbean and name it 'West Indies' than the then Pope, the Spaniard Alexander VI, granted Ferdinand and Isabella exclusive rights over the islands discovered, to the west of Azores. The Portuguese who had already obtained from the Popes unbounded rights and laid unprecedented claims over the newly discovered territories, interpreted this new papal sanction as an infringement on their rights and so remonstrated. The conflicting claims were finally defined and settled by the treaty of Tordesillas in 1494 which described the meridian lying 370 leagues west of Cape Verde Islands. The regions east of this divide were assigned to the Portuguese and the west to the Spaniards. The Pope issued a document called *Inter Caetera* in 1494 ratifying the treaty of Tordesillas.

The Portuguese wanted to tighten their authority over the area between the Cape of Good Hope and India by acquiring the prerogative to nominate the Apostolic Commissar. So they approached Pope Alexander VI who issued the bull *Cum sicut* on 26 March 1500 granting the right to the Portuguese king in view of the request made through the Cardinal of Lisbon.[49]

After the return of João da Nova from India to Portugal on 11 September 1502, the Portuguese king took a new title, *Senhor da Navegação, Conquista, e commercio da Ethiopia, Arabia, Persia, e India* and added to his existing title.[50]

The king of Portugal, Dom Manuel I known as the 'Grocer King of Europe' decided to obtain the papal sanction for the conduct of trade and commerce with the Muslims and other non-Christians and to exonerate his predecessor as well as himself from sins committed. This was thought to be indispensable for asserting the absolute monopoly of trade with the Indian Ocean regions inhabited chiefly by Muslims and Hindus. In fact all those who traded with Muslims and other non-Christians without special permission from the Pope incurred grave sin as well as the penalty of excommunication. Hence the Portuguese kings, John II and Manuel I, who established commercial relations with Muslims and Hindus in their overseas enterprises were deemed to have committed serious sins and incurred ecclesiastical censure. Accordingly, Dom Manuel supplicated before the Pope for remission of sins committed by him and his predecessors as well as his subjects. Pope Julius II listened to the petition sympathetically and issued the bull *Sedes Apostolica* on 4 July 1505, making arrangements for remission and reparation. Every one who incurred ecclesiastical censure by conducting unauthorized trade with Muslims and Hindus was given the permission to choose for himself a confessor who would recommend a suitable penance for such sins. The Pope issued an order of this nature solely in view of the fact that the Portuguese trade and commerce with the people of Guinea and India brought salutary effect to Portugal and the Church at large, and promoted salvation of souls and the glory of God. He further absolved all the sins committed by the kings of Portugal.[51] The same Pope issued another bull *Deisderas ut nobis* on 2 April 1506 permitting Dom Manuel to conduct trade with Muslims and Hindus since the desire of the king was born out of his interest for the salvation of souls.[52]

With the permission of the Popes, the Portuguese intensified their overseas activities in the East and came to clash with the

Spaniards who too were interested in attempts of this sort. Finally they had to come to terms regarding the limits of the sphere of their activities, and an agreement was reached. Dom Manuel wanted to buttress this treaty with the approval of the Pope and approached Pope Julius II who through his bull *Ea, quae* dated 24 January 1506 and ratified it.[53] Thus a clash was averted and the sphere of influence of the Portuguese was kept away from the Spaniards.

The Portuguese rulers were quite keen on seeing that their feats were brought to the notice of the Popes and took steps to make others understand that whatever they did in the East was with the sanction of the Pope. This was calculated to guard their conquests and confirm their monopoly. As soon as they conquered Malacca, the most important emporium of international maritime trade which connected the East with Venice on the east-west axis of emporia trade, this was brought to the knowledge of the Pope Leo X. The fact that Afonso de Albuquerque after his conquest of Goa brought the coveted emporium of the East under the Portuguese, was announced to the cardinals in the consistory. Through the bull *Significavit nobis* issued on 5 September 1513 the Pope thanked King Manuel of Portugal.[54] Thus Portuguese rulers always tried to show a religious touch for their activities as the pioneers of Christian faith with a view to keeping others away from interfering with their valuable acquisitions. To enhance this feature of their feats in the East, they obtained from Pope Leo X the authority to bring all the ecclesiastical institutions and benefices in the areas conquered and to be conquered from the non-Christians under the vicar of Thomar, the seat of the military order of Christ.[55]

By giving a religious tint to the overseas enterprise, the Portuguese ran the risk of the encroachment of the clerics on the jealously guarded monopoly of trade. Clerics enjoying the special privilege of not being judged or prosecuted by civil authorities, also exempt from freight charges, entered into trade and commerce with India. This was considered an infringement of the monopoly of the Portuguese king and detrimental to the finances of the kingdom. So the king petitioned the Pope for a favourable disposal of the case. Since the Pope was aware of the fact that the Portuguese trade and commerce were beneficial to the Church at large in ways more than one, he issued an order on 27 April 1521 whereby the chief chaplain of the king was given the authority to punish clerics in minor orders for offences of this type.[56]

The Portuguese who were thus protected by the Papal authority from any infringement of their all-pervading rights over the Indian Ocean regions became extremely powerful. They could dictate the price of spices in the international market. They did not have any competitors. Hence it was notified to the Popes that they were exploiting the situation and were amassing wealth through extraordinary profits in the trade in spices. Italians as well as people from the rest of Christendom made representations to the Pope to intervene in this matter. Pope Clement VII issued an order to King John III of Portugal on 9 April 1524 asking him to reduce his profits from trade in spices, and bring down the sale price.[57]

Such close interaction between the Papacy and the Portuguese kingdom emboldened clerics to derive profit from trade and at the same time cling on to the clerical privilege of not being under the secular laws. This brought about great fiscal problems for the Portuguese crown. King Sebastian approached the Pope for necessary provisions in this regard. Pope Pius IV issued a bull on 4 October 1563 admonishing the clerics and permitting the Portuguese crown to judge the clerics involved in trade and commerce without ecclesiastical authorization.[58] The privilege so far enjoyed by the clerics was done away with. The Portuguese crown benefited a lot from this in putting into effect their monopoly and plugging all possible loopholes by bringing even recalcitrant clerics under the royal orders.

Apart from taking steps to guard the newly discovered territories and maritime trade in the Indian Ocean regions from other powers with the assistance of the ecclesiastical authorities, some important steps of a secular nature were taken by Portugal. After the discovery of the sea route connecting India with the Atlantic ports and the establishment of commercial relations by Pedro Álvares Cabral, Dom Manuel, the king of Portugal augmented his grandiloquent title in 1501 by adding a new epithet, which ultimately read: 'Lord of navigation, conquest and commerce of Ethiopia, Arabia, Persia and India' (*Senhor da Navegação, Conquista e commercio da Ethiopia, Arabia, Persia e India*) as noted above.[59] Subsequently Vasco da Gama who had been to India for the second time, introduced a device whereby ships plying in the Indian Ocean regions were constrained to take a pass from the Portuguese authorities lest they be captured and confiscated by the Portuguese. This was initiated in 1502.[60] Though the payment for the pass (*cartaz*) was only nominal in the initial stages, in course of time ships equipped with passes were bound to visit the stipulated ports under the Portuguese power and pay customs duties for the items carried in the ships as it will be discussed later. Assumption of the title as well as the insistence on the obligation of *cartaz*, a humiliating one as far as the Indian merchants and elite were concerned, were important steps asserting Portuguese supremacy in the Indian Ocean.

The Portuguese claim for monopoly over trade and navigation in the Indian Ocean regions was justified in a very graphic manner by a Portuguese historian, Joao de Barros in the first half of the sixteenth century. His argument is summarized below for a better understanding of the Portuguese position. First, the assumption of the title *Senhor da Navegação, Conquista, e commercio da Ethiopia, Arabia, Persia, e India* is explained. The king rightly assumed the title in the wake of the discovery of the sea route by Vasco da Gama and especially after the return of Pedro Álvares Cabral. In fact, the king took possession of whatever was discovered and conceded and granted to him by the supreme pontiffs. This grant was made because of the large expenditure incurred by Portugal both in terms of bloodshed, sacrifice of lives of the Portuguese, and discovery undergoing sickness and dangers as well as works of thousand sorts.

Dom Manuel I, the king of Portugal took his title because the supreme Pontiffs beginning with Pope Eugene IV and Pope Nicholas V to Pope Sixtus IV granted to Portugal everything from Cape Bojador to the end of the East that would be discovered by the Portuguese. Entire India, islands, seas, ports, fisheries and so on were clearly included in the papal donations. The king ordered Vasco da Gama and Pedro Álvares Cabral to discover three things which none of the kings of Europe cared for, nor tried to discover. He wanted to take possession of these three things which were very essential in the whole of the Orient. He discovered the navigation of the unknown seas through which people travelled from Portugal to India in the east; he took possession of the route of navigation through navigating in it; the king discovered the lands inhabited by the 'idolatrous gentiles' and 'heretical Muslims' to be able to conquer and take away these lands from unlawful holders since they denied to God the glory due to Him as Creator and Redeemer. Hence the king assumed the title over these lands. He discovered the trade in spices which were dealt by those 'infidels'. Just as he was the lord of the route and of the conquest of the land it was fitting that he be the lord of trade of that land.

Barros added that there was no need for further justification for these titles other than the early apostolic grants. These titles based on the papal grant were again confirmed by the right of usucaption or prescription, as seen in the process of the Portuguese history. Here the writer refers to the undisturbed and pacific possession of the land for a stipulated period which gave proprietary rights by the legal title called prescription.

Indian mariners and traders procured certificates of safe conduct called *cartaz* from the Portuguese officials posted in India for the security of their ships. But 'infidels' from places where there were no Portuguese fortresses or with which the Portuguese had no friendly relations could be rightfully captured as in a just war.

The Portuguese writers of this period recognized the fact that seas had to be regarded as open to every one since there was no other public passage. They held that a Christian through faith and baptism was brought under the jurisdiction of Roman Church and was thereby subject to Roman law. They further held the view that the right of passage on the seas for navigators and the obligation to respect the property of those navigating in the seas were applicable only in Europe and that too only for Christians. Portugal and other kingdoms directly under the Pope observed this law not because they were subject to the imperial law as feudatories, but because these laws were just and reasonable.

Muslims and Hindus were not considered privileged to enjoy the advantages of the laws of the Portuguese. They were outside the law of Christ which was the true law. Evidently Indians lost the common right referred to above because they did not accept the Christian faith. Further, even those who received that faith were not entitled to this right because, before the Portuguese took possession of the seas, none of them had acquired any property by way of inheritance or conquest. So only the Portuguese had any right in India. This state of affairs was based on the natural principle and common law according to Barros.

With regard to the title of conquest, Sofala, Quiloa, Mombassa, Ormuz, Goa, Malacca and Moluccos with all the islands were already brought under the jurisdiction of the Portuguese king before the middle of the sixteenth century. Diu and Bassein with the lands attached to them in the kingdom of Gujarat were also in the possession of Portugal. The fortresses and their officials had been under the Portuguese king. Quiloa and Mombassa had been given up on account of sickness and lack of any good results. The Islands of Socotra and Anjediv, being not essential, were also given up. Moreover, there were other areas, the ports of which were in friendly relations with the Portuguese and received the Portuguese vessels as though Portugal was their ruling power.

The title of trade was also due to the king of Portugal. This was clear from the fact that several ships carrying spices and other sorts of commodities used to reach Portugal from India every year. Commercial relations of this nature presupposed the agreement of two contracting parties which entailed peace, friendship and observance or obedience to the contract. Beyond this general convention, Portugal had trade with India in three ways. First, trade was conducted with the conquered areas of the Indian Ocean regions by establishing commercial relations with the local people as vassals with their lords whose revenue for entry and exit belonged to the Crown of Portugal. Second, the Portuguese concluded permanent contracts with the local kings and rulers regarding the prices of commodities purchased and sold by the Portuguese as it was done with the kings of Cannanore, Chale, Cochin, Quilon and Ceylon. They were the lords of all the spices available in India. The contract was only for the supply of spices to the Portuguese officials stationed in the factories in India, for export to Portugal. Commodities other than spices could be freely purchased by privateers and local people at any price mutually agreed upon by seller and buyer. Also, the Portuguese merchant vessels plied all parts of the Indian Ocean maintaining the customs of the place and exchanging commodities with the local people at prices agreed among themselves.[61]

The monopoly claimed by the Portuguese in the light of the papal bulls and other legal presumptions prompted them to keep the entire Indian Ocean regions under their control through the introduction of passes (*cartazes*), the establishment of well-fortified fortresses in strategic places and *cafila*. Of these, the most important for and most humiliating for the local people was the system of passes.

After Pedro Álvares Cabral met with a serious rebuff from the Zamorin of Calicut from whom the Portuguese in a highhanded fashion demanded exclusive rights of trade, Vasco da Gama reached the Malabar coast in 1502 with a fleet of fifteen well-equipped ships ready for war. The rulers of Cannanore, Cochin, and Quilon were, for various reasons, sympathetic to the Portuguese, and Vasco da Gama decided to spare their ships. The Portuguese Admiral demanded that the Zamorin expel all Muslim merchants, both Indian and foreign, from Calicut: not a single Muslim should be permitted to have any relations with any port in his kingdom. There were about 5,000 Muslim families residing in Calicut itself at that time. The Zamorin stated in unequivocal terms that his port and kingdom would remain open to everyone. Fierce naval battles were thus fought in this region; the Portuguese, introduced an expedient under which those ships that were not to be attacked were required to carry a certificate signed by the Portuguese authorities (the royal factor, or the captain of the fortress). This certificate was the *cartaz* and it was for the first time issued in 1502 as mentioned above.

Later, the Portuguese officials were detailed to guard the coastal regions to prevent other ships from conducting trade with any part of India and they were asked to capture and confiscate all ships not carrying *cartazes.*[62] Merchants and rulers interested in sending their commodities to places like Ormuz, or coming to different ports of India were constrained to take *cartazes* from the Portuguese and this became a regular practice.[63] The captain of the respective fortress or the official of the factory issued the *cartazes.* Lopo Soares, the Governor of Portuguese India, issued orders in 1518 that there should be a register in which all the *cartazes* issued from time to time could be entered and was to be shown to him whenever he desired to consult it. The *cartazes* were to be made out by the writers of the factory who were entitled to a share in the perquisites, but they were to be signed by the authorities of the factory or fortress.[64] At times an amount of five pardaos for each *cartaz* was charged to the parties concerned.[65]

A *cartaz* thus issued contained the reference to the circumstance in which it was given. The name and tonnage of the ship, name and age of its captain, the port of embarkation and disembarkation, and the approximate date of departure were also indicated in the *cartaz*. Mention was made of the arms and ammunitions carried in the ship, and the items that were prohibited for transport were also declared. Lastly the names of the writers and of the issuing authority were given along with the date of issue.[66]

The insistence on the obligation of taking *cartazes* for the vessels belonging to rulers and merchants as well as the strict naval surveillance instituted to implement the system effectively caused international resistance. On account of the humiliating treatment meted out, the Zamorin of Calicut came forward to muster support for a global confrontation. He was offended by the highhandedness of the Portuguese who wanted that the merchants from Mocha, Tennasserim, Pegu, Ceylon, Turkey, Egypt, Persia, Ethiopia and Gujarat (all who frequented the port of Calicut) to be kept away.[67]

The flourishing trade brought significant revenues to the exchequer of the Zamorin. By the arrival of the Portuguese things began to take a turn altogether different. They sacked the city in 1500 and insisted on the expulsion of all the Muslim merchants from Calicut.[68] Vasco da Gama in 1502-3 demanded that the Zamorin not allow any Muslim vessel to anchor off any of his ports or to conduct any sort of trade there.[69] Since the Zamorin did not accede to the demands of the Portuguese, they bombarded the city and captured a number of vessels.[70] As a result of the frequent attacks of the Portuguese and the strict watch kept on the movement of the vessels in the Indian Ocean, the trade that had flourished for several years in the port of Calicut was dislocated; the merchants fled to other places. Besides, the Portuguese established cordial relations with the rulers of Quilon, Cochin, and Cannanore where they constructed their factories and fortresses in due course of time. This was a great blow to the prestige of the Zamorin who held sway over all the kings on the Malabar coast. He wanted to retaliate and sent envoys to Q' ansawh al-Ghawri, Sultan of Cairo, Malik Ayaz, the governor of Diu and to the ruler of Gujarat seeking help to form a united front against the Portuguese.[71]

The kingdom of Gujarat too held a pre-eminent position in trade before the arrival of the Portuguese. Many Gujarati merchants had settled in Calicut, Cannanore, and Cochin. The most important settlement was at Calicut.[72] The Gujaratis who were compared to the Italian merchants in the matter of trade in spices extended their trading operations upto Malacca in the south.[73] They took spices and other commodities to the African coast too. Now that the Portuguese began to establish their authority over the passage in the Indian Ocean and forbade other ships to navigate in this region, Gujarati vessels could not go to the ports on the Malabar coast for cargoes. Muhammad Shah, the sultan of Gujarat had appointed Malik Ayyaz as the governor of Diu who took up the cause of the Gujarati merchants. As soon as he got the complaint from the Zamorin regarding the harassment of the merchants by the Portuguese, he sent word through the merchants from Mocha to the Sultan of Cairo offering his cooperation in evicting the Portuguese from the Indian Ocean.[74]

The joint forces of the sultan of Cairo, the Zamorin, and Gujarat dealt a blow to the Portuguese in the naval battle waged in 1508 near Chaul. But retaliation by Francisco de Almeida, the Portuguese viceroy to avenge the death of his valiant son Lourenço de Almeida, reestablished the Portuguese hold. In fact, even the Venetian forces sent under the command of Amir Husayan, the governor of Jidda, were put to flight by the Portuguese. The resistence of the Afro-Asian front against the Portuguese did not thus meet with success. The Portuguese tightened their hold over the regions.[75]

The acquisition of Goa in 1510 and that of Malacca in 1511 provided the Portuguese with better opportunity to assert their exclusive claim over the Indian Ocean. These historical events made the Portuguese writers elaborate the theories regarding the hold of the Portuguese over maritime routes. Thus, the claim established by the Portuguese theoretically and practically continued to be unchallenged considerably till the fourth quarter of the sixteenth century.

The Dutch who were deprived of their right to purchase spices and other commodities after the assumption of the authority over

Portugal by Filipe II of Spain in 1580 looked for opportunities to reach the source of spices in the east. A certain Heemskerck, the captain of the employee of the Dutch East India Company, captured a large Portuguese galleon in the straits of Malacca in 1602. Some of the members who were able to distinguish between trade with the East Indies and the capture of the Portuguese vessels refused to have any share in the prize and others sold their shares in the Company. A few others decided to justify the capture of the Portuguese galleon. Hugo Grotius, a young Dutch scholar and lawyer, was called upon to bring out arguments in support of the capture. A book entitled *Mare Liberum* (the Freedom of the Seas) was anonymously published in November 1608. The fact that this tract was Chapter XII of the Treatise *De Jure Praedae* written in the winter of 1604-5 remained as a secret until 1868. The work of Grotius was written with the purpose of refuting the unjustified claims of Spain and Portugal to the high seas.

The arguments of Hugo Grotius in his *Mare Liberum* challenged point by point the claims put forward by the Portuguese for their right over the Indian Ocean coasts and especially the maritime passage. He denied that the Portuguese could have sovereignty over the East Indies on account of the title of discovery. Indian resources were famous for centuries before the arrival of the Portuguese and merchants used to come to India by sea to find the means to escape poverty; the Portuguese could not say that they had 'discovered' India.[76] Even granted that they had discovered it, it did not bring about the possibility of appropriating its wealth so long as it was not *res nullius*, or unoccupied by any one. The famous theologians like Thomas Aquinas[77] and the Spanish writer Victoria held that on the basis of religious beliefs no infidel could be deprived of his possessions and so the proprietary right of the infidels could not be taken away by the Portuguese.

As noted above, the Popes had issued a number of bulls granting the Portuguese right to the territories already discovered or to be discovered. Writers including Joao de Barros made much of the papal grants. But, Hugo Grotius very clearly denied the right of the Portuguese to eastern lands based on the papal grants. After all, the Pope did not have temporal authority and even if he did he had nothing to do with the infidels and to expropriate them in so far as they did not belong to the Church. He refuted the argument that the Portuguese had right to India on the basis of war and conquest, except in the case of Goa. The Indians being unbelievers could not be subjugated by force just because they were infidels. Nor could they be justified in saying that they fought the Indians for the sake of spreading the Christian faith, because the Portuguese were amassing wealth even to the neglect of their religious duties.

Grotius refuted also the theories regarding the exclusive right claimed by the Portuguese for dominion over the Indian Ocean since the titles such as papal donation, occupation and prescription or custom did not hold good in this regard. In the same way, claim to the monopoly of trade based on the titles of occupation, papal donation and prescription had been denied. Thus Grotius argued that the trade with India as well as the right to navigate in the Indian Ocean should not be considered a monopoly of the Portuguese but on the contrary open to all.

The English did not want to remain silent spectators. The *Mare Liberum* of Grotius was written to refute the claims of the Portuguese as well as those of the Spaniards over the Indian Ocean and the trade thereof. There was no support for English claims to the high seas to the south and east of England or to undefined regions to the north and west. William Welwood, professor of Civil Law at the University of Aberdeen, published *An Abridgement of all the Sea-Lawes* in 1613 defending the English claim. He dealt with the community and property of the seas. Two years later Welwood himself published another book in Latin entitled *De Dominio Maris Juribusque ad Dominium praecipue Spectantibus Assertio Brevis ac Methodica*. Subsequently John Selden, a famous lawyer and scholar came forward to defend the claims of the English. He wrote *Mare Clausum* in 1617 or 1618. This was aimed at refuting the arguments of Grotius and was published in 1635. In the note of dedication to King Charles I, Selden made mention of the rash attempts of foreign writers to claim the more southern and eastern sea belonging to the king of England for their princes, and to prove that all seas were 'common to the universality of mankind'. Seldon's thesis was two-fold: (a) 'The sea, by law of nature or nations is not common to all men, but capable of private dominion or property as well as the land' (b) The king of Great Britain is lord of the sea as an 'inseparable and perpetual appendant of the British Empire'. Grotius did not accept what was propounded by Welwood and prepared a reply under the title *Defensio Capitis Quinti Maris Liberi Oppugnati a Gulielmo Welwodo Juris Civilis Professore, Capite* XXVII *ejus Libri Scripti Anglica* Sermone cui Titulum Fecit Compendium Legum Maritimarum. It was considered the commentary of *De Jure Praedae*. It was published in 1872 in Muller's *Mare Clausum, Bijdrage tot de geschiedenis der rivaliteit van Engeland en Nederland in de zeventiende euw*. Once the work of Grotius was published and began to be discussed, the Portuguese took up the challenge and Frei Serafim de Freitas came up with his *De Justo Imperio Lusitanorum Asiatico* published in 1625 from Valladolid.

Freitas upheld the right of the Portuguese to the East Indies especially India on the basis of the freedom to propagate the Christian faith in non-Christian areas, thus evading the objections proposed by Grotius. He admitted that the Pope did not have direct jurisdiction over the infidels in India, but he affirmed an indirect jurisdiction. The Pope as universal pastor had the right and duty to send missionaries to the areas of unbelievers and force infidels to hear the word of God preached by the missionaries. He could also insist that an infidel ruler not prevent the preaching of the missionaries and the conversion of the people even by waging war against the ruler. Thus the Pope had indirect jurisdiction on infidels. Freitas added that the Pope who had indirect jurisdiction over the infidels and the right to send out missionaries could delegate power to any particular nation of his choice as he had in the case of the Portuguese and he was empowered to restrict the right of navigation to these areas only to a certain nation as the missionaries were to be taken by ship. Moreover, the Pope also had the power to assign the monopoly of trade to these people and to exclude others, since missionary activities needed money. This was the fundamental argument Freitas had to offer in support of the validity of papal grants to the Portuguese in regard to their jurisdiction on infidels in India, the navigation of the ocean, and the commerce with India.

Similarly, regarding the occupation of Indian Ocean which by its nature was not a thing to be occupied, he seems to have admitted the possibility of quasi-possession or quasi-occupation by reason of the privileges granted by a sovereign authority such as Pope for the sake of the activities of the propagation of faith, since he had jurisdiction all over the world. Therefore, one could acquire the right to prevent anyone from causing trouble to the peaceful navigation of the occupant of the Ocean. The explanation given for the system of *cartazes* in the Indian Ocean also was based on the papal grant, which implied that nobody else should navigate in the Indian Ocean regions without the permission of the Portuguese. As the Arabs from the very beginning of the entry of the Portuguese in the Indian Ocean tried to oust them and destroy their properties by inciting eastern rulers, it was necessary for the Portuguese to introduce such measures to curb them. This step was also necessitated by the fact that the Turks and the Egyptians were helped by Christians themselves, who lost sight of their communality of faith and fought against the Portuguese.

NOTES AND REFERENCES

1. Luís de Albuquerque, *Instruments of Navigation*, Lisboa, 1988, p. 5.
2. Ibn Khaldun, *The Muqaddimah: An Introduction to History* (tr. Franz Rosenthal), 3 vols, London, 1958, vol. I, p. 117.
3. Eleanor P. Spencer, 'L'Horloge de Sapienc. Bruxeles, Bibliothèque royale, MS, iv. iii, *Scriptorium*, vol. 17, 1963, pp. 277-99.
4. Luís Mendonça de Albuquerque, *Os Guias Náuticos de Munique e Évora*, Lisboa, 1965.
5. Joaquim Barradas de Carvalho, *La TraductionEspagnole du 'De Situ Orbis' de Pomponius Mela par Maitre Joan Fraas et les Notes Marginales de Duarte Pacheco Pereira*, Lisboa, 1974.
6. *Le Livro de Marinharia de Gaspar Moreira*, ed. Léon Bourdon and Luís de Albuquerque, Lisboa, 1977; *Livro de Marinharia de Manuel Álvares*, ed. Luís Mendonça de Albuquerque, Lisboa, 1969; Cortesão Armando et al., *Arte de Navegar de Manuel Pimentel,* Lisboa, 1969; Luís Mendonça de Albuquerque, ed., *O Livro de Marinharia de André Pires*, Lisboa, 1963.
7. Albuquerque, *Instruments of Navigation*, p. 5.
8. Francis Madison, *Medieval Scientific Instruments and the Development of Navigational Instruments in the XVth and XVIth Centuries*, Coimbra, 1969, p. 26. The illustrations on page 12 and 26 bring out the difference between the mariner's quadrant and the medieval quadrant.
9. Albuquerque, *Instruments of Navigation*, p. 6.
10. The planispheric astrolabe was not primarily an instrument for observation in the sense of determining accurately the position of a celestial body. It was, however, used observationally for telling the time by day or night, and in surveying, e.g. for determining the height of a building. In modern

terms, it is an analogue computer serving to solve astronomical problems by simulating the apparent rotation of the stars about the pole. Ref. Francis Maddison, *Medieval Scientific Instruments*, pp. 11-12.

11. Maddison, *Medieval Scientific Instruments*, p. 28.
12. João de Barros, *Da Asia,* Decada I, part I, pp. 319-20.
13. One chart is preserved in the Archivio Castiglioni , Mantua. This is dated but not signed. Three other charts by Ribero are found in Thuringische Landesbibliotheck, Weimar, 1527 (not signed) and 1529 (signed); Bibliotheca Vaticana, 1529 (signed).
14. For further details on the astrolabe, ref. Manuel Pimentel, *Arte de Navegar,* Lisboa, 1969, pp. 69-72. He defines it as 'O astralábio não é outra coisa senão um circulo de latão ou outro metal, no qual há duas travessas fixas que se cortam em ângulo recto no centro do instrumento, e uma declina ou regra móvel , a que os arábios chamam alidade, sobre o mesmo centro, com duas pinulas nas extremidades, em cada uma das quais há um buraquinho por onde entram os raios do Sol . . .', p. 69, Reis, AS. Estácio dos, *Duas Notas sobre Astrolábios*, Lisboa, 1985.
15. Instrumento de sombras, illustrated in Pedro Nunez' work contains a horizontal plate, *a b c d,* has inset a magnetic compass. In the centre of the place at *e,* there is a vertical style which casts a shadow of the sun on a scale of degree around the circumference of the place. The lines *f* and *g* show two possible positions of the shadow, and are referred to in Nunez' discussion of the use of the instrument. The compass, set in the northern half of the plate, has a south-pointing needle. Ref. Francis Madison, *Medieval Scientific Instruments*, p. 37.
16. Pedro Nunes appears to have published his first book on nautical sciences in 1537 under the title *Tratado da Sphera com a Theorica do Sol e da Lua e ho Primeiro Livro da Geographia de Claudio Ptolomeu*. Ref. Luís de Albuquerque, 'Portuguese Books on Nautical Sciences from Pedro Nunes to 1650', in Mria Emilia Madeira Santos, ed., *Luís de Albuquerque: Estudos de Historia da Ciencia Nautica, Homenagem do Instituto de Investigação Cientifica Tropical,* Lisboa, 1994, pp. 673-92.
17. Jacinto Inácio de Brito Rebelo, ed., *Livro de Marinharia. Tratado de agulha de marear de João de Lisboa*, Lisboa, 1903.
18. Ref. Francis Madison, *Medieval Scientific Instruments,* p. 47.
19. Luís de Albuquerque, *O Livro de Marinharia de André Pires*, Lisboa, 1963, pp. 133 ff.
20. Gaspar Correa, *Lendas da India*, tomo I, part I, Coimbra, 1922, pp. 261-4.
21. Luís de Albuquerque, ed., *O Livro de Marinharia de André Pires*, p. 217.
22. Joaquim Bensaude, ed., *Tratado da Esfera* (Facsimimle edition) Munich, 1915.
23. Manuel Pimental, *Arte de Navegar*, Lisboa, 1969, pp. 74-6.
24. This manuscript is preserved in the Byerische Staatsbibliothek, München.

25. Ref. *Rudimenta Mathematica, Haec in duos digerunter libros, quorum prior Geometriae tradit principia seu prima elementa, una cum rerum & uariarum figuraram dimensioni dimensionibus. Posterior vero omnigenum Horologiorum docet declinationes, autore Sebastiano Munsero*, Basle, 1551, pp. 48-50.
26. Luís de Albuquerque, *Instruments of Navigation*, Lisboa, 1988, p. 14.
27. Johann Spangeberg, *De arte mensurandi, In Compositione baculi Jacobi Ioannes Spangeberii*, published with the work of Gemma Frisius: *Gemmae Frisii, Medici et Mathematici, de Radio Astronomico &Geometrico liber*, Paris, 1558, fl. 84.
28. João de Lisboa, *Livro de Marinharia*, ed. Brito Rebelo, p. 41.
29. Albuquerque, *Instruments of Navigation*, p. 24.
30. *Ibid.*, p. 25.
31. Ramusio, *Primo Volume &Seconda editione delle navigationi et Viaggi*, Venice 1554: '. . . i marinari di la [Calicut] cioé i Mori non nauigano con la tramontana, ma com certi quadranti di legno', p. 131.
32. Pedro Nunes, *De Arte atque reatione navigandi libri duo*, fl.45r and 45v, Coimbra, 1573.
33. Manuel Pimentel, *Arte de Navegar*, p. 73.
34. *Ibid.*, p. 78.
35. Albuquerque, *Instruments of Navigation*, pp. 52-4.
36. Fernão Lopes de Castanheda, *Historia do Descobrimento e Conquista da India pelos Portugueses*, livro I, Coimbra, 1924, p. 209.
37. A. Fountoura da Costa, *Livro de Marinha de Bernardo Fernandes*, p. 112.
38. A. Fountoura da Costa, ed., *Livro de Marinharia de Bernardo Fernandes (c. 1548)*, Lisboa, 1940, p.109.
39. *Ibid.*, p. 1133.
40. *Ibid.*, p. 113.
41. *Ibid.*, p. 110.
42. 'Therefore the sea can in no way become the private property of any one, because nature not only allows but enjoins its common use; ... It is clear, therefore, to every one that he who prevents another from navigating the sea has no support in law', Donellus IV, 2. A certain Heemskerck, the captain in the service of the Dutch East India Company, captured a Portuguese galleon in the straits of Malacca in 1602 and subsequently approached Hugo Grotius, a young lawyer to justify the action. He wrote a treatise called *Mare Liberum* in 1604-5 which constituted the twelfth chapter of *De Jure Praedae. Mare Liberum* was anonymously published in November 1608 in the nature of a brief. But it remained secret until 1868 that Mare Liberum was none other than Chapter XII of *De Jure Praedae Hugo Grotius, The Freedom of the Seas or The Right which Belongs to the Dutch to Take part in the East Indian Trade,* New York, 1916, pp. 30, 44.
43. The Military Order of Christ which received papal approval in 1319 was the successor of the Military Order of the Templars established in Jerusalem

in 1119 to fight the 'infidels'. The Templars settled in Tomar not far from Santarém in Portugal in 1159. The same site was later occupied by the Military Order of Christ in 1356. If the declared aim of the Templars was to fight Muslims, the new Order of Christ was to play a significant role in the overseas and maritime expansion of Portugal with their headquarters at Tomar. Members were incorporated into this new order which was finally secularized in 1789, and continued only as on honorific order. Pope John XXII through his bull *Ad Ea* dated 15 March 1319 granted papal recognition to this new Military Order of Christ. Ref. Levy Maria Jordão, ed., *Bullarium Patronatus Portugalliae Regum in Ecclesiis Africae, Asiae atque Oceaniae* (hereafter *Bullarium*), tomo I, Olisipone, 1868, pp. 2-6.

44. *Bullarium*, p. 21.
45. *Ibid.* pp. 31-4.
46. *Ibid.* pp. 36-7.
47. *Ibid.*, p. 39.
48. *Ibid.*, pp. 47-52.
49. *Ibid.*, p. 59.
50. Joao de Barros, *Da Asia,* Decada I, part 2, Lisboa, 1777, p. 11.
51. *Ibid.*, pp. 60-1.
52. *Ibid.*, p. 73.
53. *Ibid.*, p. 70.
54. *Ibid.*, pp. 81-2.
55. *Ibid.*, pp. 98-9. Bull '*Dum Fidei*' issued by Pope Leo X on 7 January 1514.
56. *Ibid.*, pp. 122-3.
57. *Ibid.*, p. 137.
58. *Ibid.*, pp. 206-7.
59. Barros, *Da Asia,* Decada I, part 2, p. 11.
60. Gaspar Correa, *Lendas da India*, tomo 1, Coimbra, 1922, p. 298.
61. Barros, *op. cit.*, pp. 11-19.
62. *Ibid.*, Decada I, part 2, p. 21; Decada II, part I, p. 181.
63. Marino Sanuto, *I Diarii di Marino Sanuto*, tomo IV, Venice, 1881, col. 367.
64. J.H. Cunha Rivara, ed., *Arquivo Portuguez Oriental, fsciculo* V, part I, New Delhi, 1992 (rpt.), pp. 30-1.
65. Mss. Historical Archives of Goa (hereafter HAG), codex no. 3027, fol. 21.
66. *Ibid.*, codex no. 1043, fl. 50.
67. Fracansano Montalbodo, *Paesi Nouvamente Retrovati & Novo Modo da Alberto Vesputio Florentino Intitulato*, Venice, 1507, p. 94; Ludovico di Varthema, *The Travels of Ludovico di Varthema in Egypt, Syria, Arabia Deserta and Arabia Felix, in Persia, India and Ethiopia* AD *1503-08*, translated from the original edition of 1510, London, 1863, p. 151; Prospero Peragallo, ed., 'Carta del Rei D. Manuel ao Rei Catholico Narrandolhe as Viagens Portuguesas a India desde 1500 ate 1505', in *Centenario do Descobrimento da America*, Lisboa, 1892, p. 31.

68. Thomé Lopes, 'Navegação as Indias Oreintais', in *Colleção de Noticias para a Historia e Geografia das Nacoes Ultramarinas que vivem nos dominios Portugueses ou são vizinhas,* tomo II, nos 1&2, Lisboa, 1812, p. 187; João de Barros, *Da Asia,* Decada I, part 2, Lisboa, 1973, p. 49.
69. *Cronica do Descobrimento e Conquista da India Pelos Portugueses*, Coimbra, 1974, p. 33.
70. Report of Giovanni Francisco de Affaitati in Marino Sanuto, *I Diarii di Marino Sanuto 1496-1533*, tomo V, Venice, 1881, col. 129.
71. Correa, *op. cit.*, tomo I, p. 746, Fernao Lopes de Castanheda, *Historia do Descobrimento & Conquista da India pelos Portugueses,* livro II, Coimbra, 1924, 3rd edn., p. 384.
72. 'Copia de uma carta de el-rei de Portugal enviada ao Rei de Castella acerca da Viagem e successo da India', in *Centenario do Descobrimento da America*, p. 29; Duarte Barbosa, *The Book of Duarte Barbosa: An Account of the Countries Bordering on the Indian Ocean and their Inhabitants,* New Delhi, 1989 (rpt.), vol. II, pp. 86-7.
73. Tome Pires, *Suma Oriental of Tome Pires*, London, 1944, vol. I, p. 45.
74. Correa, *op. cit.*, p. 746.
75. K.S. Mathew, 'The First Mercantile Battle in the Indian Ocean: The Afro-Asian Front against the Portuguese (1508-1509)', in Luis de Albuquerque and Inacio Guerreiro, eds, *II Seminario International de Historia Indo-Portuguesa: Actas*, Lisboa, 1985, pp. 177 ff.
76. 'Quid quod ne reperisse quidem Indiam ullo modo dici possunt Lusitani, quae tot a saeculis fuerat celebrrima. iam ab Horati tempore: Impiger Extremos currit mercator ad Indos Per mare pauperiem fugiens'. Hugo Gortius, *op. cit.*, p. 12.
77. *Summa Theologica*, II-a II-ae, Questio 10, article 12.

CHAPTER VI

Life on Board a Portuguese Ship of the Sixteenth Century

The passengers and crew of Portuguese vessels coming to India and returning to Portugal constituted a microcosm. They had to be accommodated for several months on board. They had their own hierarchy of authority and values. They contracted some diseases peculiar to the tropics besides the common diseases. The mental, religious and physical aspects of their life for several months should be taken into account. Some entertainment had to be provided besides religious performance to keep the passengers under control and to keep them religious throughout the voyage taking into account the religious obligation of the country. Therefore discussions of these aspects will be offered in this chapter.

Many a scholar familiar with maritime history is currently delving deeper into various aspects of Portuguese navigation during the sixteenth century especially against the backdrop of the discovery of a direct sea-route connecting India with Iberian peninsula as well as some parts of South America. Any one going through the report of the voyage of Vasco da Gama in 1497-8 will be taken aback by the insurgence of his crew. He will equally be shocked by the strong determination evinced by the Portuguese admiral and the power wielded in quelling the rebellion. Similarly, the religious expression of the crew while facing a tempest in the sea and the psychological tensions dominating them also are quite revealing.

Social Composition and Hierarchy on Board

The social composition and the interpersonal relationships among those who manned the Portuguese vessels in the sixteenth century constitute an interesting field of study. Usually a fleet leaving Portugal for India had on an average five to six vessels. The exploratory fleet commanded by Vasco da Gama in 1497 consisted of only four vessels.[1] But the one under the same adventurer sent to Calicut to take revenge upon the Zamorin and his allies in 1502 had ten big ships and five caravels. Similarly, the fleet under the command of Francisco de Almeida, the first viceroy to India had 21 vessels. A fleet was always under a captain-in-chief who was by and large drawn from the ranks of nobility during the sixteenth century. Vasco da Gama, according to the contemporary historian Gaspar Correa, was from a family of nobles. Appointment as captain-in-chief was usually awarded to a person taking into account the services rendered by him in any field, not necessarily navigation. The captains-in-charge of the individual vessels were under the captain-in-chief of the fleet. The post of captain too was on the whole reserved to persons of noble birth and was offered to them in view of the services rendered by them in whatever field they were. This leads us to the conclusion that the captains or captain-in-chief were not persons experienced in navigation. Thus Vasco da Gama actually did not have the required experience of navigation. Only when Paulo da Gama excused himself for being the captain-in-chief on account of some previous wound was the post conferred on him. But captains had a lot of authority with the regard to the discipline in the fleet or vessel and great autonomy in relation with passengers and the crew of the vessels or fleet. A captain of a ship had the supreme authority over the ship and all those who were in it. Even the person of high nobility who might be travelling in the ship had obey him. However, when a decision of great importance had to be taken, the captain was expected to call a meeting of all the officials, noblemen, and merchants on board to sign the resolution taken in the meeting. The captain had to abide by the decision taken in such meetings. He did not have the jurisdiction to condemn any criminal to death. But he was permitted to inflict corporal punishments on board. He had the power to punish a person for civil offences and give penalty up to 200 *cruzados* without any appeal. He was allowed to keep an arrested person in chains throughout the voyage and on arrival at the destination to hand him over to a judge.[2]

A actual navigation was in the hands of the pilot, second pilot (*sotapilote*) and the rest of the subservient crew. Pero d'Alanquer, the pilot in the ship commanded by Vasco da Gama, was an experienced person who had been in the fleet of Bertholomeu Dias during the discovery of the Cape of Good Hope.[3] The pilot was the one who was sometimes seen on the top of the stern poop (*popa*), with one of the three magnetic needles (*bussalas*) found in every ship. He was often accompanied by a mariner who would transmit the message given by the pilot. He had also the obligation of writing the log book.[4] He was second only to the captain. He was always bound to watch the compass and magnetic needles. He was assisted by the second pilot (*sotapilote*). He had his cabin above in the rear part of the ship on the right side with two or three rooms. He was found always in the same cabin without moving to the hatch ways or any other area of vessels. He commanded the master of the ship to do the needful. The master under the instructions given by the pilot arranged the positions of the sails.[5]

The master of the ship (*mestre*) was next. He was directly in-charge of all the mariners (*marinheiros*), cabin boys (*grumete*), and other service personnel. The master had the responsibility of management of the ship from stern to the great mast and strike sails and all other necessary services. His cabin was just behind the cabin of the pilot on the left side of the vessel with the same number of rooms and space as the pilot. He commanded from there, with the use of a silver whistle. He looked after the main mast and its sails. Further, he was in-charge of the entire vessel and its belongings. He took care of the making and repairs of the sails with the assistance of the sailors. Similarly, any repair to be done in the vessel was executed under his care. According to need, the cannon was taken out and put back in place under instructions give by the master. Whenever he wanted cloth for sail, or nails, or rope he could obtain them from the factor of the ship or purser. He was expected to sign a receipt for them.[6]

The boatswain or foreman (*contramestre*) was appointed to help the master of the ship. He was expected to be in-charge of the area of the ship from the stem (*proa*) to the mizen mast (*maestro de mezena*) inclusively. He did in this section all what was done by the master in poop though he could not issue any orders to this effect. He controlled the work of other seamen and was in-charge of the ship's rigging, boats and anchors. The foreman had the charge of the cargo in the ship, loading and unloading the consignment on reaching the destination. His cabin was in the forecastle and he had command over the fouke mast and the fore-sails. He too had a silver whistle like the master, and took care of all things pertaining to the fouke mast. He had to look after the anchors.[7] He was not expected to leave the poop. All these personnel were appointed by the king.

There was a writer for every Portuguese vessel who was to be accountable to the king. Every thing of importance for the king as well as the individuals in the ship had to be brought to his notice, and to be registered by him. All the orders and duties of the people travelling in the ship were to be approved by him as they had some value unlike those of the French. He was bound to keep all the records related to justice in a separate office. When a crew member or passenger died, he had supposed to make an inventory of all his belongings in the ship and to dispose of them through auction to the highest bidder and also give the money on interest. On reaching the destination, a copy of the inventory had to be given to the relatives of the deceased. They were expected to defray the expenses incurred if any. The writer thus played an important role in the vessel. Nothing could take place without his nod and counsel. All victuals, even a pint of water, were distributed only with his knowledge. He kept the keys of the hatchways of the ship so much so that even when the captain wanted to go down to the cellar or stowage, the writer was expected to accompany him.

The guardian or quartermaster of the ship had a cabin close to the great mast outward on the left hand. The place on the right was used for dressing meat for the cooking. He too had a silver whistle. He had command over all the cabin boys and those who swabbed the deck with the use of pumps. He used to be accommodated along with the cabin boys during the day and late night in the space between the great mast and the mizen mast whether it rained or not. There was some leather clothing made of the hide of bull or cow to cover him. [8]

The cabin boys were under the command of the master on board the ship. If they did not come promptly at the second whistle they were beaten by sticks. They were lowest in rank among the people on board and inferior to mariners. They did not climb the masts nor did they go to the deck of the ship. They did all the hard physical labour like cleaning the ship and helped the mariners as servants; they were reprimanded or beaten up by the latter. They were bound to do work of all sorts on board and outside. They did not know how to handle the rudder of the ship.[9]

The mariners were highly respected. By and large every one of them knew to read and write because this was necessary for the art of navigation. The term 'mariner' was used to signify persons knowing everything related to navigation. The control of the ship depended entirely on them in different degrees. In big and sturdy vessels each mariner was given one or two cabin boys or help. They saw to the spreading and striking of the sails, handling cordage and other activities. But they never cleaned the vessel nor arranged guns except in urgent necessity.

Personnel were separated in to three sections during the night, one under the pilot, another under the master of the ship, and the third under the boatswain. The cabin boys with them were also divided in the same way. Each group had to keep awake for four hours at night. Every one had to be at the rudder of the ship for two hours. There were three compasses or magnetic needles in big vessels, one for the pilot on the top of the poop, another on the deck with a mariner to hear the voice of the pilot because those who went to the rudder below could not hear him. The one in the middle repeated the words of the pilot. There were two chief mariners who were called 'trinqueiros' who took care of cordage and arrangement as well as repair of sails. There were four boy-servants or pages who did not do any work other than calling people for respective work and shouted at the foot of main mast with all their strength. They called the people for specific duties, conveying messages from the master and other officials. They collected the belongings of those who died on board.

There was a bailiff (*meirinho*) or *alcaide* who executed the orders of the captain as far as administration of justice was concerned. The prisons were situated near the pump. Culprits were put here ordinarily with chains around the feet. Only the bailiff was allowed to go there. Besides, there were small prisons on a deck made of planks with holes into which the feet of men were put. The bailiff was in-charge of gunpowder, arms and ammunitions, and fire on board. There were two big kitchens with fireplaces on every deck near the mast. The bailiff lit the fire places around 8 or 9 o'clock in the evening. Two men were appointed to see that no disaster took place on account of the fire and that no one took fire to his own individual place on board. If anyone enjoying the confidence of the captain wanted to go to the cellar of the ship to see his belongings, the bailiff gave him a candle in his hand to go to the cellar after closing the exit with chains. If not, the bailiff himself accompanied him to the cellar. He had to see that the fire was put out at 4 o'clock in the morning.

The ships coming to India from Lisbon carried artisans and craftsmen of all sorts for necessary repairs: carpenters, caulkers, coopers, and others. The great part of the cabin boys were attached to them. Four of these boys were expected to sleep in the crow's nest (*cesto da gavea*) and others according to allotment. The master, boatswain, guardian or quartermaster, and the chief of the gunners had their own silver whistles hanging from silver chains round the neck to call men for specific work; the master and the boatswain to call mariners, the chief of the gunners to call all the gunners, the guardian to call the cabin boys and pages.

There were two dispensers in-charge of the stores, one for the soldiers and another for the mariners. But nothing from the stores could be distributed except in the presence of the writer of the ship. These dispensers were posted by the king.[10]

The number of people travelling in a ship of medium tonnage was between 500 and 1,000. There were approximately sixty mariners, seventy cabin boys, a chief gunner with twenty five gunners, a chaplain, a writer, four pages, a bailiff, one or two dispensers, one or two artisans or craftsmen in different fields (surgeons, carpenters, caulkers, coopers) who all told, amounted to 150. Then there were soldiers, noblemen, merchants, ecclesiasts and several men and women who were passengers of the Portuguese ship.[11] Sometimes the number of the soldiers varied from 700 to 800 in big vessels toward the end of the sixteenth century as described by Pyrard de Laval.

Tensions on Board

The social backgrounds of the people on board were varied and caused tensions. Some scholars identify the most important tensions on board arising from the relations between the merchants and navigating officers like the captain; between the navigating officers and the common sailors in the case of the East India man of the Netherlands. The strict discipline and the brutality of the commanders vis-à-vis the robust solidarity among the common sailors could pose a serious problem and create tensions.[12]

The difference of interest played a great role. Sometimes, faced with problems of overloading and unfavourable conditions of the sea, some cargo had to be thrown out. It might happen that the people would suggest this be pepper for the king rather than their own consignments. There could be tensions between the captain-in-chief and the other captains of the fleet on account of differences of opinion. Similarly, the mariners might have been developing some sorts of antagonism towards captains belonging to the nobility on land. The mariners would like to elect a captain from their own, to whom they swore allegiance. They might even reject the authority of a nobleman on board. Some times in critical moments the mariners would take a decision contrary to that favoured by men on the top of the social hierarchy so much so that the latter had to be silent. The social environment ashore was made more pronounced on board to create a wedge between the captain and the mariners. As seen earlier, the captain was normally from the nobility while the mariners were not. Moreover, there were noble men and ecclesiastical nobility on board who, according to the code of social conduct in Portugal, held the upper hand.[13]

Apart from tensions in relations between the captains and the captain-in-chief and that between the captains and the mariners, there were other factors contributing to brewing antagonism. The persons on board were separated from their families and their beloved ones for months together during their voyage. On the whole women were not allowed to make voyages, especially in this male-dominated society for a long time. Instances of sodomy, a crime inviting capital punishment according to the moral contact of the time were not uncommon. Presence of the few lone women either clandestinely or legally especially in case of a new governor, could be the focus of tension in a male dominated group on board.

Despite the prohibition for Portuguese women to travel to India, some were found on board ships leaving Portugal for India. Thus many women were found in a ship coming to India in 1548 as reported by the Jesuit fathers.[14] Further, some women even of bad reputation clandestinely travelled on the ships. The Jesuit fathers travelling in the ships for India identified some and took sufficient precautions to keep them under control. Thus a letter written by a certain Antonius de Quadros, SJ on 18 December 1555 at Goa made mention of a woman of doubtful fame discovered on board. She was shifted to another ship with the consent of its captain-in-chief and shut up in a room with a couple of watchmen. On reaching India, she was sent to a house of married women in Goa.[15] The Jesuits used to check the ships at the time of departure from Lisbon for such women stowaways. Thus, a certain Fernandes da Cunha on 16 March 1562 before setting sail for Lisbon, checked thoroughly the ship to see if any women were found on board.[16]

Fear was another important factor that disturbed the people on board; fear of death, and fear of travel by sea in general. A vivid and picturesque description about agony and anguish of the death evinced on the persons on board the ship has been given by Dom Gonçalo in his letter of 1557 from Cochin. He says that the feelings of those who approached death were beyond description. They seemed not to have heard about death any time in their life. The limbs were sometimes frozen on account of the wind blowing at the Cape of Good Hope.[17] When the co-passengers saw several of their friends succumbing to death, their own agony and anguish was heightened. The moment they reached a port, they shouted for joy and gladness. As soon as a storm approached passengers panicked, some began to pray and even went to confess. The pilots and the physicians had to spend a lot of time consoling people on board. It has been appropriately said by the same priest that people on board died several deaths till they reached their destination.

Diet on Board

Before the ships left Lisbon for India, they were fully provided with necessary victuals. The problem of managing food for such a group of men on board a Portuguese ship that took several months to reach its destination, without facilities for refrigeration, was complicated. The tremendous change of climate when the ship passed through the tropical regions could badly affect the stores. Sometimes on account of contrary winds the vessels were bound to spend several days near Equador where food spoiled on account of the heat. It was reported by the passengers that greater part of edible and potable items like butter, oil, marmalade, water, figs, raisins and honey were spoiled on account of the hot weather.[18] The same was the case with salted meat that got completely spoiled near the coast of Guinea on account of being soaked.[19]

Water was very essential for the life on board and potable water was always a rare and expensive item in the ships. Moreover, it used to get spoiled fast on account of the climatic changes on the way from Lisbon to India and back. Travellers used to mention the adverse effect of heat experienced on the African coast that used to spoil potable water. Sometimes the passengers had to shut the eyes and close their nostrils to drink water provided on board especially near Moçambique, on account of the high degree of pollution as reported by an eyewitness.[20]

Water was very carefully distributed daily. Sometimes to get extra water, one had to pay a lot. It was reported that twenty days after departure from Lisbon, potable water became scanty and expensive – so much so that a unit of 32 litres of water (*almude*) was sold for 480 *reis* in 1567.[21] On account of the importance of water-management, an inspector from among the priests, a certain Estevão Lopes was appointed to look after its distribution and to take special care not to waste it.[22] Each passenger had to be satisfied with two and a half ports of drinking water a day. It was insisted that nobody should go for more water than ordered. On account of the scarcity and lack of water of good quality, some compared it with the water they used to have at Évora and even prayed to God to give them better water. Attempts were made to have stopovers, where good water could be fetched for the life on board, but on account of the dependence of the voyage on monsoon winds it was sometimes difficult to reach the desired stopover or hauling station. Hence, it was suggested to collect rain water and store it for drinking purposes by fixing necessary devices.

With a view to avoiding decomposition of victuals on board, living beings that could be used for food were allowed to be transported on board the ships. Thus fowls, castrated rams, pigs, sheep, and roosters were taken on board. Cocks were useful also in ascertaining the time based on their morning singing.[23] Sometimes the prices of these living beings shot up beyond imagination just as in the case of water. The price of a fowl reached 500 *maravedis.*[22] There were often complaints that even by paying 1,000 *reales* it was difficult to buy a fowl on board.[24]

Bread could be obtained as and when the people on board wanted from the respective dispenser according to need. Wine and water were distributed daily according to a stipulated measurement namely, 2.622 litres (half a *Canada)* per person.[25] Salted meat were supplied once a month. Approximately 13.11 kg. of meat were the ration per person for a month. Oil, vinegar, salt, onion and fish were also distributed in the same proportion on a monthly basis. All this was done in the presence of the writer who took account of every thing. If a person did not drink wine he could sell it to another or he could return it to the dispenser in exchange for anyother item. Those who had victuals for sale could sell them at whatever price they wanted.

Fish was a much sought-after item for those on board, especially when they spent many days without fresh food. There were many fishing areas on the way from Lisbon to India and back. The zone around the Cape of Good Hope abounded in fish of various species. Portuguese mariners used to catch fish in this area in abundance over this stretch of the voyage.[26] The victuals supplied in raw form had to be cooked by each individual according to his needs, which was considered detestable by some passengers. At a time there were 80 to 100 pans on fire since each person was expected to cook for himself.[27] In fact the sick passengers were put to trouble since there was no common cook. A Dutch priest, Gaspar Braz by name on board a Portuguese vessel in 1948 complained that there was no one to cook for him while he was sick. All colleagues were busy cooking for themselves and at last he had to follow them. Even the cabin boys did not help him. The difficulty continued until a certain Henrique Macedo instructed his own servant to cook for this priest.[28] Of course the nobility and the officials on board had their own cooks from among their slaves or servants. On some festive occasions all passengers were given food cooked by the cabin boys and others employed in the ship. But this was not a regular phenomenon.[29]

Health Hazards and Care

A number of people on board became sick for various reasons. The change in climate, food habits, the background from which those on board were recruited and a host of other factors could be the ground for sickness of various sorts. It was reported that many used to become sick on the way.[30] Most of the sickness appeared after the ships crossed the zone of the Guinea coast. Long voyages coupled with lack of fresh food and necessary vitamins like C could be some of the reasons for common sickness found among the people on board. The sick were generally looked after by the missionaries, chiefly Jesuits on board, though the medical profession was forbidden to them. The missionaries, though they were fully qualified doctors, had to go to barbers of the ship for bleeding.[31]

The most common sickness found among the people on board was scurvy, known as *escorbuto, mal de Loanda, hidropsia escorbutica, mal dos marinheiros* or *enfermidade do edema.*[32] The well-known descriptions of the symptoms of the sickness are found in the Diary of the first voyage of Vasco da Gama attributed to Alvaro Belho. The gums of those who were affected by this disease swelled and grew over the teeth in such a way that they were not able to eat anything. Similarly, the legs and feet were also swollen. There were swellings all over the body so much so that person affected by this malady suffered till they died of it. It was reported that thirty persons died of this in approximately three months' time in the first voyage besides many others earlier.[33] João de Barros makes mention of the same disease contracted by the mariners under Vasco da Gama when they were for a month on the African coast. Many people became sick and a few died of this malady. Their gums swelled as if they could not be accommodated in the mouth. The flesh decomposed.[34] Luis de Camões while talking about the difficulties faced by the mariners makes mention of the disease in his famous *Lusiadas.*[35]

Scurvy continued to afflict the people on board for a long time. Oranges, a source of vitamin C, were supposed to be great remedy for scurvy. Vasco da Gama himself managed to get large quantity of oranges from Melinde for the people who were sick on board.[36] Moçambique was the most scurvy-prone area in the *Carreira da*

India. Oranges always constituted an important prophylactic item in the treatment of scurvy.

Nausea (*enjoo*) was another malady of the people on board. It recurred whenever the sea was rough. The travellers on board *Santiago* in 1585 were badly affected by this when the sea was very rough.[37] Since there was no appropriate remedy available for various sicknesses, several of the people on board died. Constant requests were made to the king to send a physician out on every ship.[38] Another disease that affected the people on board very commonly was petechial fever (*tabardilho* or *brentaeja*). This was like an epidemic. The first documented evidence of this disease on board was that of the three ships, *S. Martinho, S. João* and *Santa Maria* which left Lisbon on 5 April 1597.[39] There were other minor diseases peculiar to the people on board.

Every ship had its boutiques containing a few remedies for sickness contracted by the travellers. Sometimes the contractors of the victuals were bounded to keep boutiques on board provided with necessary medicines. There was often scarcity of qualified medical practitioners on board the ship especially for tropical diseases. The most common remedy suggested by the medical practitioners was the treatment by surgeons who assisted in making the sick bleed or using purgatives. Sick people were often asked to go to the barber or surgeon (*cirurgião*). It was reported that with one or two bleedings could affect a cure.[40] Sometimes a patient suffering from tropical diseases had to undergo five bleeding as reported by co-passengers.[41]

The Care of the Sick

In his first voyage to India St. Francis Xavier started looking after the sick. A number of people on board the ship, which left Lisbon on 7 April 1541, became sick especially after they crossed the coast of Guinea. So he and his companions took charge of the sick and helped them die well.[67] The care for the sick on board was willingly taken up by the Company of Jesus from the time of Francis Xavier. A lot of people died on board the ship en route to India and back of diverse diseases. Brother Jacome de Braga of the Company of Jesus described in 1563 the appalling condition of the sick on board. He says that three to four persons died every day on account of sickness.

They were so sick that they could not eat. As soon as someone became sick, he was advised to go for the sacrament of Penitence to keep soul purified.[68] Right from the time the ships set sail a number of people on board would become sick.[69] The condition of the sick became aggravated for lack of qualified doctors and nurses. Hence, the fathers did whatever they could, even by providing them better meals cooked by them.[70] The Jesuit Fathers who travelled to India from Lisbon played the role of nurses and dispensers of medicines and victuals. In fact they were not allowed to follow the profession of doctors since it was forbidden to them in the Canon Law. At times they suggested that there should be a person appointed as the chief of those who took care of the sick. The Jesuits seeing the necessity of some special arrangement for the sick, suggested that a chief physician be appointed to every vessel coming from Portugal and that they would willingly take care of the sick. They offered to provide the sick with special food, or else clothing or a bed. Those sick that were not properly treated were given special care by the Jesuits. Even those who took care of the sick fell prey to the sickness. The number of the sick went on increasing. Many people were put to bleeding by the barbers .The captain stored a lot of oranges in the ship to give to the sick.[71] The ordinary mariners always suffered from the lack of food. They depended on the goodwill of the officials, noblemen and the priests on board.[72]

A report of 28 November 1574 speaks of Father Vallone taking care of the sick. Later Father Peter Ramon took this up with great enthusiasm. His services were jealously watched by others. Some of them went to the extent of cooking meat ad serving it to the poor and the sick on board.[73] Bleeding was often resorted to for almost all diseases. It was believed that by permitting the malignant blood to ooze out from the veins, sickness would be cured. St. Francis Xavier himself had undergone bleeding seven times during his journey from Lisbon to India.[74]

The Religion of those on Board

In the face of indomitable nature and the vagaries of weather especially in the tropical regions, people had recourse to religion to a degree unknown to those on land during the sixteenth century.

Every Portuguese ship had a captain and a priest on board for religious services for those who travelled in it. They were paid by the king. The priest was bound to celebrate the holy mass every day including holy days. But the Blessed Sacrament was not allowed to be preserved in the vessel. He had the obligation to hear confessions, give homily and to perform all ecclesiastical functions. Besides, other religious persons used to travel in those vessels in India without having any such obligation against payment though they performed religious duties voluntarily.[42] Some of the sacramentals of the catholic religion too were at the disposal of those who wanted to lead a pious life on board. Holy water was preserved in vessels. Lent, feast days, and processions were observed on board. The oratories and the chapels in the vessels were beautifully decorated with panels of framed pictures giving a congenial atmosphere for prayer. When someone passed away on board, the master of the ship informed others about it. Through his whistle he called them to pray for the repose of the soul. But there were no shots as was usual on lands. Every night at nine, the master invited everybody for prayer through his whistle. All those on board as a group recited *Our Father* and one *Hail Mary* in unison, after which at the sound of another whistle by the master they retired to his post or quarters to fulfil their own obligation to pray. Similarly at dawn the cabin boys intoned a hymn which was repeated by all those on board. Every one in the subsequent prayer made mention of the ship, and all the things in it. This prayer lasted for a good hour. Everyone prayed in a loud voice.[43]

The life of prayer found on board the Portuguese ship was enviable. It was remarked by a passenger that the adage that if anyone wanted to learn to pray, he should enter a ship, seemed to him factual.[44] A lot of activities of piety were performed on board. Several factors were considered to be responsible for an active religious life. The most important fact that the voyage started from Lisbon usually in the month of March and continued through Lent during which every conscientious Christian was supposed to perform a lot of pious activities. This was further accentuated by the presence of priests and the religious on board who used to preach the Gospel, singing hymns in honour of Blessed Virgin Mary, the patroness of voyages, led processions, and heard confessions. Moreover there was a conviction among the people that the success of a voyage, to a great extent, depended on the purity of life on board. Jesuit fathers thus drove away two women of doubtful character from among the persons on board to avert punishment from God.[45] This seems to be a conviction shared by many coastal people, especially fishermen, all over the world.[46] A life of piety was necessary to avoid violent tempests, epidemics, and God's punishment.

Invariably a voyage started during Lent which demanded a life of atonement for transgressions of religious principles in the past and revitalization of a life based on moral injunctions. Confessions and other religious rituals were very sedulously observed. Palm Sunday, Maundy Thursday, Good Friday, Easter, and the Sunday in white during the paschal season were very carefully and solemnly celebrated. Washing of feet, processions, discussion on the passion of Christ and the singing of hymns were conducted on Maundy Thursday under the leadership of the Jesuits in the fleet of 1562. Confessions of a number of people on board were heard by the priests on board.[47] It must be remembered that faced with a tempest in the sea and prospective disaster several people went to confession as a preparation for a good death.

On the Palm Sunday the priest on board the ship conducted the accustomed ceremonies of blessing the palms, reading a passion narrative, and the celebration of solemn mass. Instructions on the observations of the holy week were given to the people as usual.[48] Everyday seven psalms or rosary was recited in groups. Sermons were so heart-rending that the people in large numbers went to confessions with deep felt repentance. Sometimes when a person on board was sick, he was asked to go for confession and bleeding, and to prepare his testament so as to have a clean conscience.[49] Bodily cure should be preceded by the purification of the soul.[50]

Entertainment on Board

Though the people on board did not have a lot of time for entertainment and were busy with various activities, there were still some theatrical performances. Some of these were meant for their edification and expression of piety during the season of Lent. The passion plays conducted during Lent were meant to remind the passengers of the suffering and the death of Christ.[51] A hard life on board coupled with anxiety, physiological tension and fear from the enemies prompted them to follow a very strict religious life. The earliest mention of such a theatrical performance is found in a letter written by Father Bartholomeu Vallone on 28 November 1574.[52] The performance was organized on the feast of *Corpus Christi*. Another drama on the life of St. Barbara was staged on the sector after the Cape of Good Hope. There were thus a number of performances related to saints, to Christ, and various aspects of Christian life.

Auctions were also conducted occasionally. As soon as a passenger died, his belongings were auctioned. Similarly after a lot of fishing on the way, there was auction. These auctions also kept people on board busy and entertained.

It was reported by the close of the sixteenth century that approximately 1,500 soldiers were sent from Portugal to India every year out of which about 1 per cent returned to Portugal. Of the rest, some died of natural causes, some were murdered, some could not return to Portugal on account of poverty. Therefore the last category of soldiers stayed in India though they did not want to. If by chance any soldiers returned, it was with the viceroys or nobles who were allowed to take a few soldiers with them on their return to Lisbon. So, twenty to thirty solders were also accommodated on board the ships. Besides, those who held an important office in Portuguese India were allowed to take a few slaves with them when they returned to Lisbon. Therefore they too were accommodated on board a ship returning to Lisbon.[53]

Spiritual Service

Preaching the Gospel on board a ship for the consolidation of faith was one of the important services rendered by the members of the society of Jesus during the voyage from Lisbon to India. Even during the stopover in Moçambique on the East African coast en route to India, preaching the Gospel, penitential services, preparation of the severely sick people for death and distribution of holy communion to crew and passengers enabling them to obtain plenary indulgences granted by Popes were some of the routine services performed by Master Xavier and his companions. A letter written in Spanish on

1 January 1542 at Moçambique by Francis Xavier gives the details of such activities.[54] A sixteenth century dictum was: 'If you want to learn to pray, enter into sea'.[55] The most important factor that made the people on board interested in prayer was the imminent and likely danger of death, for which they should be prepared. They had to be ready to accept it without hesitation. Another circumstance was the Lenten season during which half of the voyage was conducted. There were orders from the king regarding the presence of priests on board to satisfy the spiritual needs of the passengers. Besides, the conviction that the success of the voyage depended to a large extent on a life of purity, instilled into the hearts of the passengers by the accompanying clergy, made them pray earnestly to God.

The missionaries on board carried with them the necessary articles of liturgical and paraliturgical services.[56] They carried even the branches of olive from Portugal for the ceremonies of Palm Sunday. After the celebration of the holy mass in which all the people on board participated, a feast used to be organized to the satisfaction of all the passengers, crew and the officials. We have an instance of a solemn celebration of the Palm Sunday on board on 22 March 1962 described graphically by Fernand da Cunha from Bassein.[57] Similarly, the washing of the feet, processions and other religious ceremonies were performed on board very meticulously by the members of the Company of Jesus.[58] Effective sermons were preached on Good Friday along with the chanting of the lamentations and solemn divine office. Young boys were invited to sing hymns and in some cases even profane songs to keep the people instructed in religion and also entertained. Vespers were also sung. In fact, the Company of Jesus did everything in board to win the people for Christ in view of the words of St.Paul to the Corinthians.[59] At times when the winds were unfavourable and a disaster was in the offing, the Jesuit Fathers gave spiritual exhortation to the people on board to embolden them to face the inclement weather. The crew and the passengers became very well disposed after giving up blasphemies and perjury.[60]

The practice of conducting solemn processions continued to be in vogue. Very often processions were organized in thanksgiving for surviving violent tempests or for getting favourable wind for the voyage, deliverance from pestilence or sickness. Crucifix and relics of saints were carried in the processions organized on board. Solemn chanting of hymns, Litany (*Ladinha*) and special prayers were offered through the intercession of the Blessed Virgin Mary.[61] Trumpets used to add solemnity to the processions and chants.[62] Hail Marys and Salve Regina were solemnly sung often. Sometimes the religious celebrations on board exceeded those in the parishes back home in their solemnity. The pilot was also involved in such celebrations.[63] Sometimes three of the fathers of the Company of Jesus conducted services of Litany in three different places simultaneously on board the same ship. We have Father Peter Ramon, father Anon Veles and Father Vallone performing such sacraments for the people on board in 1563.

The Jesuits sometimes converted some of the people on board to Christianity. A report dated 28 November 1574 refers to the conversion of a Muslim on board. He was seriously sick. Several times they asked him about his willingness to become Christian. Always he refused. Later they commended him to God and prayed for his conversion. Once again they asked him. He immediately accepted Christianity and recovered.

Administration of the sacrament of penance was an important aspect of the life on board. Many of the people on board were in great existential anguish about their life whenever the sea became rough and some of their colleagues died of some disease. Therefore many of them confessed their sins and got absolution from the priests on board. St. Francis Xavier writing on 1 January 1542 states that on the way from Lisbon to India, he was busy with the confessions and distribution of Holy Communion.[64]

A report refers to the administration of the sacrament of baptism for an infant born of a married woman on board. Father Peter Ramon baptized the child. Later the child died and so the last rites were performed for it. Three persons, who were helped for spiritual and temporal needs by the Fathers, were given the sacrament of the sick. When they died on board the father performed the last rites for them on board.[65]

The Fathers of the Company went on deepening their knowledge of theology and sacred scripture on their way to India. Experienced and learned persons from among themselves gave lectures on theology and the New Testament to others. They performed, besides, several spiritual exercises for themselves on board.[66]

Discipline

The society on board the caravels of the *Carreira da India* was a male-dominated one. Taking into account the long duration of the voyage, and the possible hazards of life, women by and large were prevented from travelling to India in the sixteenth century. Yet Father Gaspar Barzeus says that there were many women of dubious character on board the ship and the Jesuits asked the captain to send them out as soon as the ship anchored near the coast.[75] Some of them were advised by the Jesuits to give up their bad ways after making a good confession.[76] They tried to keep the people at peace with one another and made them forgive the offences committed by others on board. The captain and other officials of the ships of the India run became quite satisfied with the activities of the Fathers of the Company of Jesus since their presence and spiritual care helped keep discipline on board. In certain cases, to keep discipline on board, the few women on board were kept in isolation. Sometimes at the time of setting sail from Lisbon, the women, especially those of dubious character who were found on board were asked to leave. In 1555 a woman of bad repute was found in one of the ships. She was briefly removed to the ship in which some of the members of the Company were travelling. Later, on account of some inconveniences, the captain-in-chief made arrangements for her in the ship *Assunção*[77] and a special chamber was set apart for her. She was locked up there. When she reached Goa, she was taken to the house of a married woman and was put there so that she might amend her way of life.[78] The Company of Jesus was very active in getting rid of women of suspicious character from the ships. This attempt on their side continued to be in vogue throughout the century. Thus, in 1562, they drove away two women of dubious character at the very time of departure from Lisbon. A certain Brother Vicencio took great interest in contacting the captain to see that women of such nature were sent away.[79]

Theatrical Performance

Voyage from Lisbon to India took eight to ten months on the average and also almost the same duration for the return. The people on board needed always some entertainments to get rid of boredom. The Company of Jesus devised ways and means to inculcate into their hearts religious ideas and ideals through theatrical performance that served the double purpose. We have the earliest reference to such a performance in a letter written by Father Bartolomeu Vallone SJ at Bassein on 28 November 1574.[80] He makes mention of a couple of religious dramas played on board the ship *Santa Barbara* which left Lisbon for India on 21 March 1574 along with four other ships, namely *Chagas, Fé, Anunciada* and *Santa Catharina.*[81] Easter fell on 11 April 1574. A theatrical performance was staged on 12 April, the day after Easter. Father Vallone composed this play in Portuguese. This was known as the *Dialogo das Três Marias*. This was to represent the visit of the sepulchre in the liturgy. The three Marys were Maria Salome, Maria Cleophas and Maria Magdalena. They while approaching the sepulchre of Jesus Christ asked who would remove the stone at the entrance of the sepulchre. A performance of this type in the liturgy was conducted in the Cathedral of Ruan in the thirteenth century. The Biblical event of the appearance of the angel dressed in white garment was also presented here.[82] Another performance was conducted on the feast of *Corpus Christi* which was celebrated on 10 June 1574. A solemn candlelight procession was conducted on the day as if it would look like a procession in a great city. Afterwards, the drama composed in Spanish by Father Peter Ramon of the Company of Jesus was staged.

During the voyage from the Cape of Good Hope to India, the comedy of Sta. Barbara was enacted. Father Vallone composed the script in Portuguese. This was liked by all the people on board so much so that they suggested that it could have been conducted even in a city. The patroness of the ship was Sta. Barbara who suffered martyrdom. The script was about the life and death of Sta. Barbara. Diogo Sanches of Badajoz (+1549) had earlier written a farce on Sta. Barbara.

One more drama, the script of which was written in Portuguese by Father Bartholomeu Vallone, was staged on board the same ship. It was about the miracle of Our Lady (*Dialogus Miraculi Dominae Nostrae*).The *dramatis personae* were Blessed Virgin Mary, Jesus Christ, a sinner and anorted devils. It was so appealing that even the superior of the members of the Society of Jesus on board shed tears.[83]

Another drama was staged on board on the occasion of Pentecost. Flavius Gregorius in his letter dated 3 December 1583 writes about it on board the ship *St. Francisco* sailing to India. This was called *Imperador*. The Portuguese had the custom of electing a young man of lower status as the emperor for a short period. People of even higher status were expected to serve him and pay obeisance.[84] The morale of this was to show that the glory of this world lasts for a short time. The governors appointed to India had usually a term of three years and they were to take this into account. Hence, according to the practice in Portugal, on 29 May 1583, on the day of the Pentecost, this play was staged on board the ship *St. Francisco*. A boy was elected as the Emperor on the vigil of Pentecost. He was clad in costly garments and an imperial crown was placed on his head. A few noblemen were elected to serve him as officials and assistants. The captain of the ship was appointed as the administrator of his house. All the officials of the ship were enlisted as people to assist him in one way or other. They prepared an altar in the prominent and spacious place on board. The Emperor elect was escorted with fanfare to the altar and was seated on a velvet chair with cushions. He wore the crown and held the sceptre. Gun salutes were also arranged during the mass. Later a great banquet was organized in which even the nobles had to serve him. Approximately 300 people on board paid obeisance to him. Another drama on the life and death of St. John the Baptist was also staged on the same day on board the ship. A Portuguese wrote this.[85]

To conclude, a ship coming to India from Portugal during the sixteenth century was a microcosm in which persons of different ranks and professions were found. Each one had a special duty to perform. It was not devoid of the tensions found in any other society and perhaps more pronounced on account of the environment. The people were alert to the fact of death and some of them took a positive step towards this universal phenomenon while others cherished a negative attitude. The innumerable religious practices actively participated by many speak for the anguish of death and their pended on the religious. The charitable works performed by the religious were highly appreciated by the sick. In fine, one may say that the life on board a Portuguese ship during the sixteenth century differed a lot from that on the indigenous ships on account of the duration of navigation and the social background of the people.

NOTES AND REFERENCES

1. According João de Barros, they were *S. Gabriel* commanded by Vasco da Gama, *S. Rafael* commanded by Paulo da Gama, the eldest brother of Vasco da Gama, *Berrio* under Nicolão Coelho, and the *Nao* commanded by Gonçalo Nunes. Ref. Joao de Barros, *Da Asia*, Decada I, part I, Lisboa, 1778, p. 279. Castanheda makes mention of four vessels namely *S. Gabriel* of 120 tons commanded by Vasco da Gama, *S. Rafael* of 100 tons under Paulo da Gama, *Caravel Berrio* of 50 under Nicolão Coelho, and another vessel carrying victuals under Gonçalo Nunes. He adds that the caravel got the name *Barrio* from the pilot with the same name. Fernao Lopes de Castanheda, *Historia da Descobrimento e Conquista da India Pelos portugueses*, Coimbra, 1924, livro I, part I, pp. 8-9. Caspar Correa gives the names of the three vessels as *S. Rafael* commanded by Vasco da Gama, *S. Gabriel* commanded by Paulo da Gama, *S. Miguel* under Nicolao Coelho. Obviously there is confusion of names here. Gasper Correa, *Lendas da India*, tomo I, part I, Coimbra, 1922, p. 15. He says that each ship carried approximately 80 passengers. He adds that Vasco da Gama was of noble lineage. He does not mention the fourth vessel.
2. *Viagem de Francisco Pyrard de Laval, Contendo a Noticia de sua navegação as Indias Orientais, Ilhas* de *Maldiva, Maldiva,Maluco e ao Brasil, e os differentes casos que lhe aconteceram na mesma viagem nos dez anos que andou nestes paises (1601-1611) com a descricao exacta dos costumes, leis, usos, policia e governo: do Trato e Comercio que neles ha: dos Animais Arvores frutas doutras singularidades que ali se encontram*, Porto, 1944, vol. 2, pp. 142-3.
3. Barros, *op. cit.*, p. 279; Castanheda, p. 9.
4. Francois Pyrard de Laval, vol. 2, p. 143.
5. John Huyghen van Linschoten, *The Voyage of John Huyghen van Linschoten to the East Indies*, London, 1875, rpt. Delhi, vol. II, p. 231.
6. *Ibid.*, p. 231.
7. *Ibid.*
8. Laval, *op. cit.*, p. 144.
9. *Ibid.*, p. 141.
10. *Ibid.*, p. 146.
11. Francisco Contente Domingues e Inacio Guerreiro, *A Vida a Bordo la*

Carreira da India (seculo XVI). Instituto de Investigação cientifica Trotical, Lisboa, 1988, p. 198; Laval, *op. cit.*, p. 142.

12. Kren Degryse, 'Social Conditions and Tensions on Board the Eighteenth Century East India Ships', in K.S Mathew, ed., *Mariners, Merchants, and Oceans: Studies in Maritime History,* New Delhi, 1995, pp. 341-6. While dwelling on the grounds of tensions between the merchants and the naval officers of Dutch ships, Degryse says that the supercargos or merchants of the East India Company were appointed by the Company and were the real leaders of the expedition while the captain did not enjoy the prestige of the supercargos. The captain did not have any jurisdiction over the merchants though as far as navigation was concerned he held the upper hand. At times the captain belonged to a different nationality. Another source of tension identified by him was the differences between the navigating officers and the common sailors. The captain had to maintain order and discipline on the ship and he usually ruled with a strong fist. The punishments meted out to sailors were out of all proportion. Against the harsh discipline by the captain there could be strong solidarity among the common sailors. Degryse gives an example of sodomical relation between two sailors who were put in chains for the rest of voyage, sodomy being capital crime to be judged by the higher court. Some of the accomplices helped each other to escape the ship when it anchored. We come across prisons in Portuguese ships for criminals.
13. 'Once on a shore, one of the first acts of *homens do mar* was to elect a captain from among their own and to whom they swore loyalty. It was common place for the mariners to follow their own captain and reject the authority of the *fidalgo* or other leader of the landsmen. Here it was the forces of social conformity, and the imposition of sanctions by the nobles or pseudo-gentry ceased to apply. The seamen emerged as sub-centre, adopting their own norms of behaviour, values and rules of conduct. By virtue of their access to resources, officers and mariners exerted a disproportionate amount of influence in determining the election of a leader in taking decisions, acting in all cases out of their own self interest as a group rather than in the collective interest of the survivors. Such was their dominance that nobles and priest were forced to remain silent in the face of blatant assaults on prevailing Portuguese codes of behavior and values as *homens do mar* commanded vessels, shared limited rations and resources between their own, and took decisions involving the life and death of unfurnates who were thrown over board on skuffs.... Social environment ashore was replicated afloat with but minor modifications and that social distinctions based on birth and rank were not only preserved but were all the more in evidence'. A.J. Russel-Wood, 'Men under Stress: The Social Environment of the *Carreira da India*, 1550-1750', in Luis de Albuquerque and Inacio

Guerreiro, eds., *II Seminario Internacional de Histiria Indo-Portuguesa: Actas*, Lisboa, 1985, pp. 19-35.

14. Joseph Wicki, ed., *Documenta Indica,* vol. I (*1540-1549*), Rome, 1948, p. 384.
15. *Ibid.*, vol. III (*1553-1557*), Rome, 1954, p. 387.
16. *Ibid.*, vol. V (*1561-1563*), Rome, 1958, pp. 569-70.
17. A. Silva Rego, ed., *Documentação para a História das Missões do Padorado Portuguese do Oriente*, vol. VI, Lisboa, 1951, pp. 188-90. 'Deixando a recordacao da navegaçào que fizemos desse para este outro mundo, porque assi como a morte nao a pinta senao quem more, nem se pode ser pintada senâo vendo quem está morrenda,assi o trago que passâo os que navegam de Portugal a India,nâo o pode cntar senâo quem o passa nem o pode enterder senâo que o ve passar. E assi, como os homens que primeira vez se viram na hora da morte, Ihes parece que nunqua oviram de faller nella, assi quem se vio em aquelles golfâos nâo Ihe alembrava cousa que Ihe tivessem ditto de verdade e terror presente,e sua,que passada, nem bem imaginar se pode. Assi,e sem maise nem menos, a angustita e agonia em que se vem os passageiros desde occidente ao oriente, em que estamos os que nos vemos for a dela nunqua a podemos vivamente representar comnosco mesmos,. . . . Nunqua se virâo suores de morte como os que se suâo na Costa da Guine Nunqua se virâo membros frios como os que cortâo os ventos de Boa Esperança... Nunqua se vio morrer homem cercado de termores e saudades do que neste mundo deixa e no spera, como os que se vem nesta carriera, vendo muytos mortos e landçados ao mar e todos os outros, antre os quais ficam velhos ainda para morrer de fome, de sede, de doenças, gravisimmas e de perigos do mar innumaraveis, de boxos, de penedos, de costas, de encontros de naos e de servo de mares. . . . De modo que se pode dizer que tantas vezes morrem os que fazem esta viajem, quantos pontos de morte vem claramente que ande passar, tendo tâo provado ficar em algum delles'
18. Joseph Wicki, ed., *Documenta Indica*, vol. VI (*1563-1566*), Rome, 1960, p. 35.
19. *Ibid.*, vol. VIII (*1569-1573*), Rome, 1964, p. 279.
20. *Ibid.*, vol. VI (*1563-1566*), p. 228.
21. *Ibid.*, vol. VII (*1566-1569*), Rome, 1962, pp. 364-5.
22. *Ibid.*, vol. VIII (*1569-1573*), Rome, 1962, pp. 228.
23. *Ibid.*, vol. VI, p. 38. The price of a hen went up to 500 *maravedis* in 1563. *Maravedi* was an old Gothic coin used in Portugal and Spain.
24. Joseph Wicki, ed., *Documenta Indica*, vol. VI, p. 383.
25. *Canada* was the unit of ancient measurement of liquids in Portugal which was equal to 2.622 litres. 300 *Canadas* made a *pipa*.
26. Joseph Wicki, ed., *Documenta Indica*, vol. XI (*1577-1580*), Rome, 1970, pp. 757-8.
27. Pyrard de Laval who travelled on a Portuguese ship reminds the readers that this practice was rather burdensome and that the French and Dutch

navigators had common cooks for those on board. He adds that at a time six person ate from the same plate on board these vessels. Pyrard de Laval, *op. cit.*, pp. 147-8.

28. *Ibid.*, vol. I (*1540-1549*), pp. 383-4.
29. *Ibid.*, vol. V (*1561-1563*), p. 217.
30. *Ibid.*, vol. VI (*1563-1566*), p. 772.
31. *Ibid.*, vol. VII (*1566-1569*), p. 280.
32. Maria Benedita Almeida Araujo, *Enfermidades e Medicamentos nas Naus Portugueses* (Seculos XVI-XVIII), Lisboa, 1993, pp. 2-3.
33. *Diario da Viagem de Vasco da Gama Fac-simile do Codico original transcri çâo e versâo em Grafia actualizada com uma introduçao por Damâo Peres,* Porto, 1945, p. 112.
34. Joâo de Barros, *Da Asia,* Decada I, part 1, Lisboa, 1778, p. 291.
35. E foi que,doença crya e feia
    A mais que eu nunca vi, desemparam
    Muitos a vida, e em terra estranha e alheia
    Os ossos para sempre sepultaram.
    Quem haverá que,sem o ver, o creia,
    Que tâo disformamente ali lhe incharam
    As gingivas na boca, que crescia
    A carne e juntamente apordrecia?
    Apordrecia c'um fetido e bruto
    Cheiro, que o ar vizinho inficionava
    Nâo tinhamos ali medico astuto
    Cirurgiâo subtil menos se achava;
    Mas qualquer, neste oficio pouco instruto,
    Pela carne jâ podre assim cortava
    Como se for a morta, a bem convinha,
    Pois que morto ficava quem a tiha.
    Ref. Luís de Camôes, *Os Lusiadas,* Lisboa, 1972, p. 159 (Canto V).
36. *Diario da Viagem de Vasco da Gama,* vol. II, p. 115.
37. Bernardo Gomes de Brito, *Historia Tragico- Maritima em que se escrevem chronologicamente os naufragios que tiverâo as Naos de Portugal, depois que se poz em exercicio a Navegaçâo da India,* vol. II, Lisboa, 1736, p. 436.
38. Maria Benedita Almeida Araujo, 'Enfermidades e Medicamentos nas Naus Portugueses', unpublished, p. 10.
39. Quirino da Fonseca, ed., *Diarios da Navegação da carreira da India nos anos de 1595, 1596, 1597, 1600 e 1603*, Manucripto da Academia das Sciencias de Lisboa, 1938.
40. Joseph Wicki, ed., *Documenta Indica,* vol. VII, p. 280.
41. *Ibid.*, vol. VI (*1563-1566*), p. 38. For more information on surgeon on board, ref. José Rodrigues de Abreu, *Luz de Cirugioens Embarcadissos, que Trata das*

Doenças epidemicas, de que constumão enfermar ordinariamente todos os que se embarcão para as partes ultramarinas, Lisboa, 1711.

42. Laval, *op. cit.*, p. 142.
43. *Ibid.*, pp. 149-50.
44. 'Se queres saber a Deos rogar, emtra em o mar' ref. Joseph Wicki, ed., *Documenta Indica*, vol. VI, p. 306.
45. Joseph Wicki, ed., *Documenta Indica*, vol. V, p. 530.
46. Coastal society in South India shared the view that the women who stayed at home should lead a life of purity to bring safety to the persons going to fish at sea. The case of Karuthamma and Parekkutty in the famous malayalam novel, *Chemmeen*, may be recalled here in this connection.
47. Joseph Wicki, ed., *Documenta Indica*, vol. V, pp. 530-1.
48. *Ibid.*, vol. V, pp. 572-3.
49. *As Gavetas da Torre do Tombo*, vol. V, Lisboa, 1965, p. 357 .
50. Joseph Wicki, ed., *Documenta Indica*, vol. VIII, p. 284.
51. Marino Martins, *Teatro Quinhentista Nas naus da India,* ed. Broteria, Lisboa, 1973, is an example of studies of this nature.
52. Martins, *ibid.*, p. 15.
53. John Huyghen van Linschoten, *The Voyage of John Huyghen van Linschoten to the Indies*, vol. 2, London, 1885, p. 230.
54. Schurhammer and Wicki, *Documenta Indica*, vol. I, Rome, 1948, pp. 91-3.
55. Joseph Wicki, *Documenta Indica*, vol. VI, Rome, 1960, p. 306.
56. *Ibid.*, vol. V, p. 572.
57. *Ibid.*, vol. V, pp. 568-80.
58. *Ibid.*, vol. V, p. 530.
59. St.Paul to Corinthians, I, Chapter 9, versicles 20-1.
60. Joseph Wicki, *op. cit.*, vol. II, p. 241
61. *Ibid.*, vol. III, p. 110.
62. *Ibid.*, vol. III, p. 109
63. *Ibid.*, vol. VI, pp. 34.
64. Schurhammer and Wicki, *op. cit.*, vol. I, p. 91.
65. Joseph Wicki, *op. cit.*, vol. IX, p. 458.
66. *Ibid.*, vol. IX, p. 237.
67. Schurhammer and Wicki, *op. cit.*, vol. I, p. 91.
68. Joseph Wicki, *op. cit.*, vol.VI, pp. 56-57.
69. *Ibid.*, vol. VI, p. 772.
70. *Ibid.*, vol. I, p. 384.
71. *Ibid.*, vol. VI, pp. 772-5.
72. *Ibid.*, vol. V, p. 57.
73. *Ibid.*, vol. IX, p. 458.
74. *Ibid.*, vol. IX, p. 93.
75. *Ibid.*, vol. I, p. 384.
76. *Ibid.*, vol. I, pp. 388-9.

77. *Ibid.*, vol. III, p. 387.
78. *Ibid.*, vol. III, p. 387.
79. *Ibid.*, vol. V, p. 530.
80. Martins, *op. cit.*, pp. 15-16.
81. Joseph Wicki, *op. cit.,* vol. IX, pp. 451-9.
82. Martins, *op. cit.*, pp. 22-35.
83. Joseph Wicki, *op. cit.*, vol. IX, p. 457.
84. *Ibid.*, vol. III, p. 103.
85. *Ibid.*, vol. XII, p. 881.

CHAPTER VII

Conclusion

As large part of the world is being covered with water, passage through the aquatic zone turns out to be a necessity for day to day life especially for commercial, cultural and religious purposes of every sort. Religion and culture, besides trade and commerce spread, beyond all the natural confines. This was impossible without transportation through high seas. Therefore the art of building ocean going vessels or naval architecture as well as the techniques of navigation developed from time immemorial. Various sorts of vessels with diverse materials were built in different parts of the world taking into account the climatic condition of the high seas and also the nature of transportation. Consequently vessels of different shapes, tonnage and material have been found in the past all over the world. As the nature and purposes of voyage through the seas and the oceans differed, ocean-going vessels too underwent structural differences from time to time. Those involved in building ships needed guidelines from experienced shipwrights in view of the wisdom accumulated by them through years or centuries. Transmission of knowledge of this nature assumed different forms, such as written communication or oral tradition handed down from generation to generation.

Indian shipwrights did not hand down written guidelines or plans for building ships. This was presumably because their children who were to be the future shipwrights actively took part in the various stages of shipbuilding from early childhood Even the experienced carpenter did not prepare any plan for the construction of a house or building of a ship. There was no separate institution to train future carpenters or shipwrights. As with any other profession, carpentry was the expertise of a particular social group in the caste based society of India. Therefore there existed no need for a separate institution to train carpenters. In other words, there are no written or oral forms of guidelines for the techniques of building ships.

On the other hand, we come across a few indigenous works in verse related to the selection of suitable timber and even to the launching of vessels. It is against this backdrop that a few works like *Kulatturayan Kappalpattu, Kalavattupattu, Kappalsattiram, Navoi-sattiram,* and *Kannakiyum Cheermakkavum* have been identified. Though *Yuktikalpataru* and a few Sanskrit sources speak of the selection of timber and types of ships, recent scholars are not inclined to take them as scientific works.

I have been able to bring to light a few European sources related to shipbuilding written in the sixteenth century. These works delve into the technicalities of building ships with detailed designs and measurement. They have been written by experts who had a very vast exposure to the art of shipbuilding in various parts of the world, such as Fernando Oliveira. When we go further into detail we come to know that they were not the people involved in building ships, but they were just scholars who provided practical guidelines to carpenters or shipwrights. Fernando Oliveira who wrote two highly sophisticated works in Latin and Portuguese in the last quarter of the sixteenth century, was a Catholic priest of the Dominican order. European society did not have the division of labour based on caste or professional groups as in India. This seems to be point of significance when we make a comparative study of shipbuilding in India before the arrival of the Portuguese and after.

Similarly, there is a great difference in the case of the guidelines for navigation too. Indian navigators did not leave any documented instructions regarding navigation. A few *Rehmanis* and *Roznamas* of later period have been brought to light, and a certain *Thangal* family of Kavratti possesses some of these materials. While there are a few Arabic and Turkish guidelines for navigation in the Indian Ocean regions, Indian mariners handed down information regarding navigation from generation to generation They were traditionally navigators and presumably did not find it necessary to write down the guidelines. They and their predecessors plied in the same region of the Indian Ocean for centuries. But the Europeans after crossing the Cape of Good Hope and entering into the Indian Ocean regions came across an Ocean entirely new to them. Besides, the Portuguese and other European navigators were employees of the Crown or the Companies. Portuguese sources, especially the *roteiros* and *regimentos* composed by the persons who actually went aboard ships contain practical instructions for the navigators. Those who exposed themselves to the unaccustomed milieu of the Indian Ocean regions found it necessary to jot down the differential aspects of navigation in this area. As mentioned earlier, these works have been not composed by persons involved in the work of navigation or mariners but by those who were observers or knowledgeable persons entrusted with the task of preparing the *roteiros* and *regimentos.* Here too we find differences in Indian and European sources.

The quality of timber available on the western coast of India surpassed that of other regions in India or even that of Europe. The temperature of the Indian Ocean differed from that of the Atlantic and so too requirements of timber for building ships. The western coast of India was blessed with the availability of teak and *angeli* which had stupendous qualities to ward off shipworms. These trees had bitter and resinous sap essential for keeping the shipworm away. They were strong and useful for masts. There were a number of other trees which were suited for various other parts of the ship, the hull, ribs, and so on. Calicut for centuries together occupied the pride of place in shipbuilding on account of the easy transportation facilities through river from the forested areas and areas suitable for the treatment of the timber for ships. The importance of Beypore near Calicut could not be challenged by any other part on the western coast of India till the Portuguese established themselves in Cochin. Similarly the shipbuilding centres in Bassein commanded great importance. Ships were built on the western coast of India chiefly for trade and commerce without the use of arms and ammunitions. Canons were not usually mounted on such vessels. This had a great impact on the regional technology of building ships before the arrival of the Europeans.

Indian navigators drawing on the wisdom accumulated through centuries of navigation by their ancestors plied in the Indian Ocean either in connection with trade and commerce or religious purposes, freely during the period before the arrival of the Portuguese. Nobody placed restrictions of any sort on their navigation. Taking advantage of the clear sky and the brightness of the celestial bodies especially the constellations, and the sun and moon, they were able to locate the target they set before setting sail. Natural energy coupled with the human energy provided propulsion for their vessels. The instruments used for fixing the location of their vessels and measuring the distances were unsophisticated. Navigation in the period prior to the Portuguese arrival was unrestricted though the specialization of a group of professionals unlike in the case of the Portuguese and of Europeans in general.

The Portuguese, determined to extract the fabulous profit of the seaborne trade and commerce with the East, arrived on the western coast of India by the end of the fifteenth century after a number of attempts to discover a direct route from the ports of the Atlantic Ocean. They established their installations like the factories and fortresses mostly on the confluence of the rivers joining the sea, to have better access to the areas in the hinterland producing pepper and other commodities of their interest. Places like Calicut, Cochin, Cannanore, Quilon, Cranganore, Old Goa, Chaul, Bassein, Diu and Daman where they opened factories and fortresses were of great importance as far as maritime activities were concerned, strategically and navigationally. The Portuguese concluded treaties and signed contracts with local rulers to conduct trade and commerce. In certain cases they put up stiff fight with some of them like the Zamorin of Calicut and Bahadur Shah of Gujarat. They extended their influences and contact not only on the coast of the Arabian sea but also the Bay of Bengal. Taking into account the importance of setting up installations for careening and building ships in India they established centres of shipbuilding where they had friendly relations with the local rulers and also where there were natural facilities.

The Portuguese took into confidence the local carpenters to harness their experience in felling the suitable trees in right season. They used some for indigenous techniques for caulking and embalming their vessels. They gave up the practice of using wooden nails in vogue on the Malabar coast and used imported iron nails. As the tonnage of the ships rose in tune with the export of larger volumes of commodities, this change became inevitable.

Many canons as well as arms and ammunition had to be mounted on the ships on account of the competition from other west European powers towards the end of the sixteenth and the beginning of the seventeenth centuries. This necessitated changes in the naval architecture. Similarly the status of the persons travelling from India to Portugal and vice versa compelled the Portuguese to enhance the standard of on-board facilities. Arrangements were also provided for religious practices on board.

The Portuguese, because of the problems faced at Calicut from time to time, set up a centre of shipbuilding at Cochin on a firm footing. This turned out to be the major installation of shipbuilding on the Malabar coast causing lasting harm to the centuries-old shipbuilding at Beypore near Calicut. They managed to get suitable timber like teak, *angeli* and other materials from the interior places through the backwaters. The fatal blow dealt to the shipbuilding industry at Beypore could not be remedied even in the later period.

Portuguese mariners unaccustomed with the waters of the Indian Ocean prepared a number of *roteiros* and *regimentos* for navigation. They were not prepared by the navigators themselves, but by the people appointed for the same. Initially they tried the indigenous equipments like *kamal* from which they made their own versions. They followed astronomical navigation after observing the changed situation in the Indian Ocean regions and adopted some of the principles of navigation from this region. The compass used in Europe was also used by them to find the direction while they were on the high seas. Presumably this was an innovation in Asian waters. They employed a number of new instruments of navigation as we have seen. The Portuguese made use of slaves and those captured in naval confrontations who were kept in the galleys for the propulsion of ships. This was not a practice in India.

Another aspect of Portuguese navigation which created a lot of resentment among Indian rulers, merchants and navigators was the ascertion of armed and exclusive rights of navigation. This was something unheard of in the Indian waters. The Portuguese system of passes was detested by all those were accustomed to free navigation for centuries together. This claim on exclusive navigation was based on the papal authority, questionable as far this area was concerned. Untill 1502 the Indian rulers and merchants had enjoyed the freedom to send their ships anywhere in the Indian Ocean regions. There were few naval confrontations except probably some attacks from pirates.

Life on board the ship of the India run of the Portuguese differed a lot from that of the Indians and Arabs in the Indian Ocean regions. The composition of the society on board consisted of mariners, officials some times including the viceroys or governors, soldiers, religious people (either ordained priests or even candidates for priesthood in the religious congregations like the Jesuits, and chaplain/s) slaves in the galleys, captain, pilot and various others who manned the ship, and very rarely women. When the ship entered into a certain zone, invariably the people on board contracted diseases which were peculiar to tropical regions. Appropriate treatment of such diseases was not often available with the physicians on board. A number of them died and their bodies were buried in the high seas which in its turn caused a lot of panic to others who were infected with diseases of various nature. The service of generous men was highly appreciated. The Jesuits came forward to help their brethren. They also helped to the captain to maintain some discipline among those on board. Some of them besides the chaplains performed religious services proper to the religious seasons like the lent. Confessions, religious processions, theatrical performances suited to the seasons were common on board the Portuguese ship. When we compare this with the life of the people on board the Indian or Arab ships in the Indian Ocean regions, this aspect of the Portuguese navigation stands out as totally distinct.

In fine one may say that when we consider various aspects of Indian shipbuilding and navigation vis-à-vis the Portuguese maritime activities in India during the sixteenth century, there are substantial points of departure as well as assimilation or accommodation. Indian shipwrights and navigators were quite advanced in their technology possibly because of the fact that shipbuilding remained as a profession traditionally taken up by a particular social group. Even without any plan drawn in advance they were able to bring out sea-worthy vessels for long distance voyages. Portuguese shipwrights on the other hand depended on plans or designs prepared by experts who were actually not shipbuilders. The timber used in the Atlantic was not generally suitable for the voyages in the Indian Ocean regions and so in course of time the Portuguese came to depend on Indian wood both on account of its availability in plenty in India and also the scarcity of appropriate timber in Portugal. In the milieu created by the maritime activities of the Portuguese, Beypore near Calicut the internationally reputed centre for shipbuilding and navigation was relegated to the background and in its place Cochin emerged as an important centre of shipbuilding. Beypore could not recover its position even after a couple of centuries .On the other hand, Cochin became a well equipped modern shipyard in the twentieth century in the wake of new techniques of shipbuilding.

Bibliography

Albuquerque, Luís de, *Escalas da Carreira da Índia*, Lisboa, 1978.

———, *Contribuição das Navegações do sec.xvi para o conhecimento do Magnetismo terrestre*, Coimbra, 1970.

———, *Estudos de História da Ciência Nautica*, Lisboa, 1994.

———, *Instruments of Navigation*, Lisboa, 1988.

———, *Navegação Astronomica*, Lisboa, 1988.

———, *O Livro de Marinharia de Pero Vaz Fragoso*, Lisboa, 1977.

———, *Um exemplo de 'cartas de Serviços'da Índia*, Coimbra, 1979.

———, *Duas Obras inéditas do Padre Francisco da Costa*, Coimbra, 1970.

———, *Portuguese Books on Nautical Science from Pedro Nunes to 1650,* Lisboa, 1985.

Alves, Francisco and Filipe Castro, 'Recent Discoveries of Portuguese Shipwrecks: New Missing links in the origin of the Iberian-Atlantic Shipbuilding Tradition', in Lotika Varadarajan (ed.), *Indo-Portuguese Encounters: Journeys in Science, Technology and Culture,* vol. 1, New Delhi, 2006, pp. 351-70.

Apte, B.K., *A History of the Maratha Navy and Merchantships*, Bombay, 1973.

Araujo, Maria Benedita Almeida, 'Enfermidades e medicamentos nas Naus Portugueses (Séculos XVI-XVII)', Lisboa (unpublished).

Arunachalam, B. (ed.), *Chola Navigation Package*, Mumbai, 2002.

——— (ed.), *Essays in Maritime History*, vol. 1, Mumbai, 1998.

——— (ed.), *Essays in Maritime History*, vol. 2, Mumbai, 2002.

———, *Heritage of Indian Sea Navigation*, Mumbai, 2002.

———, 'Indigenous Traditions of Indian Navigation with Special Reference to South India', in K.S. Mathew (ed.), *Studies in Maritime History*, Pondicherry, 1990, pp. 127-42.

———, 'Indigenous Traditions of Indian Navigation' – Report of the CSIR Project, 1995.

———, 'The Haven-Finding Art in Indian Navigational Traditions and Cartography', in Satish Chandra (ed.), *The Indian Ocean: Explorations in History, Commerce and Politics,* New Delhi, 1987, pp. 191-221.

———, 'Timber Traditions in Indian Boat Technology', in K.S. Mathew (ed.), *Shipbuilding and Navigation in the Indian Ocean Region A.D. 1400-1800,* Delhi, 1997, pp. 12-19.

———, 'Traditional Sea and Sky Wisdom of Indian Seamen and their Practical Applications', in Himanshu Prabha Ray and Jean-François Salles (eds.), *Tradition and Archaeology: Early Maritime Contacts in the Indian Ocean*, New Delhi, 1996, pp. 261-81.

———, 'Traditions and Problems of Indian Nautical Cartography', in Victor Rajamanickam and Y. Subbarayalu (eds.), *History of Traditional Navigation Thanjavur,* 1988, pp. 93-106.

———,'The Chola Mode of Navigation in the Northern Indian Ocean', in Lotika Varadarajan, ed., *Indo-Portuguese Encounters: Journeys in Science, Technology and Culture,* vol. II, New Delhi, 2006, pp. 417-37.

———, *Navigational Hazards, Landmarks and Early Charting: Special Study of Konkan and South Gujarat*, Mumbai, 2007.

Barata, J.da G. Pimentel, *O Tratado das Naus e Galeões Portgueses de 1550-80 a 1640*, Lisboa, 1970.

———, *A Ars Nautica do Padre Fernando Oliveira–Enciclopedia de Conhecimentos maritimos e primeiro tratado cientifico de construção naval (1570)*, Lisboa, 1970.

———, *Estudos de Arqueologia Naval* (2 vols.), Imprensa Nacional Casa de Moeda, Lisboa, 1989.

Barata, Jaime Martins, *O Navio 'São Gabriele' e as Naus Manuelinas* , Coimbra, 1970.

Barata, José Alberto Leitão, 'Os Senhores da Navegação, Aspectos nauticos do Dominio Portugues no Oriente por metados do século XVI'. unpublished M. Phil dissertation, University of Lisboa, 1992.

Barbosa, Duarte, *The Book of Duarte Barbosa*, 2 vols., Nendeln / Liechtenstein, 1967.

Barcellos, Constantino, 'Construção de naus em Lisboa e Goa', *Boletim da Sociedade de Geografia de Lisboa,* Lisboa, 17ª serie, 1898-9.

Barros, Eugenio Estanislaus de, *As Gales Portugueses do século xvi*, Lisboa, 1930.

———, *Traçado e Construção das Naus Portuguesas dos séculos XVI e XVII,* Lisboa, 1933.

Barros, João de, *Decadas da Asia,* Lisboa, 1777, 4 Decadas.

Blot, Jean-Yves, 'The Chinese Junk: Archaeology of a Dream', *Review of Culture*, Macao, 1997, pp. 201-50.

Bourdon, Léon and Luís de Albuquerque, eds., *Le Livro de Marinharia de Gaspar Moreira,* Lisboa, 1977.

Bowen, Richard Le Baron, 'The Dhow Sailor', *The American Neptune*, vol. XI (1951), pp. 161-203.

———, 'Arab Dhows of Eastern Arabia', *The American Neptune*, vol. IX (1949), pp. 87-132.

Bulhão Pato, Raymundo António de (ed.), *Cartas de Affonso de Albuquerque*, tomo II, 1898; tomo III, 1903; tomo IV, Lisboa.

Calado, Adelino de Almeida, *Livro que trata das cousas da India e do Japão*, Coimbra, 1957

Castanheda, Fernão Lopes de, *Historia do Descobrimento e Conquista da India pelos Portugueses*, Coimbra, 1924-33, 9 vols.

Castro, Felipe, 'The Remains of a Portuguese Indiaman Carrack at Tagus Mouth', in *Proceedings of the International Symposium on Archaeology of Medieval and Modern Ships of Iberian –Atlantic Tradition'*, Lisboa, 1998, pp. 79-87.

Castro, João de, *Roteiro de Goa a Dio (1538-39)*, O Porto, 1843.

———, *Roteiro de Lisboa a Goa (1538)*, Lisboa, 1940.

Chandera, C.M.S., *Kannakiyum Cheerammakkavum*, Kottayam, 1973.

Chandra, Satish, B. Arunachalam and V. Suryanarayanan, *The Indian Ocean and its Islands-Strategic, Scientific and Historical Perspectives*, New Delhi, 1993.

Chaudhuri, K.N., *Trade and Civilization in the Indian Ocean: An Economic History from the Rise of Islam to 1750*, Delhi, 1985.

Chitins, Ajay H., 'Navigation in the Indian Ocean: A Comparative Study between European and Indian Maritime Techniques in the Medieval Period', in *Proceedings of the Twentieth Anniversary Seminar on European Influence in the Shaping of Maritime India*, Mumbai, Maritime History Society, April 1999, pp. 21-8.

Cipolla, Carlo M., *Guns and Sails in the Early Phase of European Expansion 1400-1700*, London, 1965.

Congreve, Captain H., *A Brief Notice of Some Contrivances Practised by the Native Mariners of the Coromandel Coast in Navigating, Sailing and Repairing their Vessels* also Garbriel Ferrand, *Introduction a l'Astronomie Nautique Arabe*, Paris, 1928, pp. 24-30.

Correa, Gaspar, *Lendas da India*, Coimbra, 1922, tomos 4.

Cortesão, Armando, *Contribution of the Portuguese to Scientific Navigation and Cartography*, Coimbra, 1974.

———, *Nautical Science and the Renaissance*, Coimbra, 1974.

Cortesão, Armando, Fernanda Aleixo and Luís de Albuquerque, eds., *Arte de Navegar de Manuel Pimentel*, Lisboa, 1969.

Costa, A. Fountoura da, *A Marinharia dos Descobrimentos*, Lisboa, 1939.

———, *Roteiros Portugueses ineditos de Carreira da India de seculo XVI*, Lisboa, 1940.

Couto, Diogo de, *Decadas da Asia*, Lisboa, 1778, 8 Decadas.

Crone, Ernst, *How did the Navigator Determine the Speed of his Ship and the Distance Run?*, Coimbra, 1969.

Degryse, Karel, 'Social Conditions and Tensions on Board the Eighteenth Century East India Ships', in K.S. Mathew (ed.), *Mariners, Merchants and Oceans*, Delhi, 1995, pp. 341-6.

Domingues, Francisco Contente, e Inácio Guerreiro, *A Vida a Bordo na Carreira da Índia* (Século XVI), Lisboa, 1988.

———, *Experiencia e Conhecimento na Construção naval Portuguesa do seculo XVI e os tratados de Fernando Oliveira*, Lisboa, 1985.

———, *Os Navios do Mar Oceano: Teoria e empiria na arquitectura naval portuguesa dos séculos xvi e xvii*, Lisboa, 2004.

Elbel, Martin Malcolm, *The Portuguese Caravel and European Shipbuilding: Phases of Development and Diversity*, Lisboa, 1985.

Falcão, Luis Figueredo de, ed., *Livro em que se contem toda a Fazenda e Real Patrimonio dos Reinos de India Portuguesa e ilhas adjacentes (1612)*, Lisboa, 1859.

Faria, Francisco, Leite de and Avelino Teixeira da Mota, *Novidades Nauticas e Ultramarinas numa informação dada em Venzia em 1517*, Lisboa, 1977.

Fatimi, S.Q., 'History of the Development of Kamal', in Himanshu Prabha Ray and Jean-François Salles (eds.), *Tradition and Archaeology: Early Maritime Contacts in the Indian Ocean*, New Delhi, 1996, pp. 283-92.

Fernandez, Manuel, *Livro de Traças de Carpintaria*, Transcição e tradução em Inglês, Lisboa, 1995.

——— (1616) *Livro de Traças de Cartpintaria, Facsimile of the Manuscript of the Library of the Ajuda*, MS. No. 52-xvi-21, Academia de Marinha, Lisboa, 1989.

Fonseca, Querino de, *Diario de Navegações da Carreira da India nos annos 1595, 1597, 1600 e 1603*, Lisboa, 1938.

Gardiner, Robert (ed.), *Conway's History of the Ship: The Age of Galley: Mediterranean Oared Vessels Since Pre-classic Times*, London, 1995.

Gopal, Lallanji, 'Art of Shipbuilding and Navigation in Ancient India', *Journal of Indian History*, vol. 40 (1962), pp. 313-28.

Gorakshkar, Sadashiv and Kalpana Desai, *The Maritime Heritage of India*, Bombay, 1989.

Greeshmalatha, A.P., 'The Battleships of Medieval Kerala', in K.K.N. Kurup (ed.), *India's Naval Traditions (The Role of Kunhali Marakkars)*, Delhi, 1997, pp. 82-6.

Greeshmalatha, A.P. and G. Victor Rajamanickam, 'The Race-boats of Kerala and their Tradition: Some Observations', in K.S. Mathew (ed.), *Shipbuilding and Navigation in the Indian Ocean Region AD 1400-1800*, Delhi, 1997, pp. 55-61.

———, 'The Ship-building Technology as Practised in Beypur, Kerala', in K.S. Mathew (ed.), *Shipbuilding and Navigation in the Indian Ocean Region* AD *1400-1800*, Delhi, 1997, pp. 44-54.

Gunawardana, R.A.L.H, 'Changing Patterns of Navigation in the Indian Ocean and their Impact on Pre-colonial Sri Lanka', in Satish Chandra (ed.), *The Indian Ocean: Explorations in History, Commerce and Politics*, New Delhi, 1987, pp. 54-89.

Hariharan, K.V., 'Sea-dangers in Early Indian Shipping', *Journal of Indian History*, vol. XXXIV, pp. 313-20.

———, 'Some Aspects of Ancient Shipbuilding and Navigation', *Journal of the University of Bombay,* 34, parts 1-4 (July 1963-January 1966), pp. 26-42.

Hornell, J., 'The Origin and Ethnological Significance of Indian Boat Designs', *Memoirs of the Asiatic Society of Bengal,* 7(1918/23).

———, 'The Tongue and Groove Seam of Gujerat Boat Builders', *Mariner's Mirror*, 16 (1930).

Hourani, G.F., *Arab Seafaring in the Indian Ocean in Ancient and Early Medieval Times,* Princeton, 1951.

Hutter, Lucy Maffei, *A Madeira do Brasil na Construção e Reparos de Embacações*, Lisboa, 1985.

Iria, Alberto, *Da Navegação Portuguesa njo Ìndico no século XVII*, Lisboa, 1973.

Karani, R. and B. Arunachalam, 'Traditions of Shipbuilding in Lakshadweep and Minicoy', in Satish Chandra, B. Arunachalam and V. Suryanarayan (eds.), *The Indian Ocean and its Islands: Strategic, Scientific and Historical Perspectives*, Delhi, 1993.

Kentley, Eric, 'The Sewn Boats of India's East Coast', in Himanshu Prabha Ray and Jean-François Salles (eds.), *Tradition and Archaeology: Early Maritime Contacts in the Indian Ocean,* Delhi, 1996, pp. 247-60.

Khader, C.K. Abdul, 'Navigation and Ship-Building on the Malabar Coast (1400-1600)', unpublished M.Phil dissertation submitted to Pondicherry University, 1993.

King, David A., *Astronomy for Landlubbers and Navigator: The Case of the Islamic Middle Ages*, Lisboa, 1984.

Kulkarni, A.R., 'Shipbuilding, Navigation and Maritime Activities on the West Coast of India in Medieval Period with Reference to the Maratha Power', in K.S. Mathew (ed.), *Shipbuilding and Navigation in the Indian Ocean Region* AD *1400-1800*, Delhi: Munshiram Manoharlal, 1997, pp. 1-11.

Kumar, K. Prem, 'The Influence of Europeans in the Development of Ports and Shipbuilding Industry in India', in *Proceedings of the Twentieth*

Anniversary Seminar on European Influence in the Shaping of Maritime India, Mumbai, Maritime History Society, April 1999, pp. 71-

Kurup, K.K.N., 'Indigenous Navigation and Shipbuilding on the Malabar Coast', in K.S. Mathew (ed.), *Shipbuilding and Navigation in the Indian Ocean Region AD 1400-1800,* Delhi, 1997, pp. 20-5.

———, *India's Naval Traditions*, New Delhi, 1996.

Lamb, Ursula, *Nautical Scientists and Their Clients in Iberia (1508-1624)*, Lisboa, 1984.

Lane, Frederic, C. 'Venetian Shipping during the Commercial Revolution', in F.C. Lane, *Venice and History*, Baltimore, 1966, pp. 1-24.

Lanman, Jonathan T., *Life on a Portuguese Nao: Linschoten's Voyage to India 1583*, Lisboa, 1985.

Laval, François Pyrard, *The Voyage of François Pyrard of Laval to the East Indies, the Maldives and Brazil,* Delhi, 2000.

Lavanha, João Baptista (1608-16), *Livro Primeiro da Architectura Naval. Facsimile, transcrição e tradução em inglês do Manuscrito da Real Academia de la Historia de Madrid, Colecção Salazar, codice 63*, Academia de Marinha, Lisboa, 1996.

Lenz, Walter, 'Voyages of Admiral Zheng He before Columbus', in K.S. Mathew (ed.), *Shipbuilding and Navigation in the Indian Ocean Region AD 1400-1800*, Delhi, 1997, pp. 147-54.

Linschoten, John Huyghen van, *The Voyage of John Huyghen van Linschoten to the East Indies,* 2 vols, Delhi, 1988.

Madison, Francis, *Medieval Scientific Instruments and the Development of Navigational Instruments in the XVth and XVIth centuries*, Coimbra, 1969.

Martins, Xavier Mariona, 'Portuguese Shipping and Shipbuilding in Goa 1510-1780, unpublished Ph.D. thesis, Goa University, 1994.

Mathew, K.M., *History of the Portuguese Navigation in India,* Delhi, 1988.

Mathew, K.S., ed., *Shipbuilding and Navigation in the Indian Ocean Region AD 1400-1800,* Delhi, 1997.

Mathew, K.S., 'Ship Building on the Malabar Coast and the Portuguese during the 16th & 17th Centuries', in T. Jamal Mohammed (ed.), *Perspectives on Kerala Traders*, Sree Sankaracharya University of Sanskrit, Kalady,

———, 'The Jesuits and the Health Services on Board the Ships of India Run (Carreira da India) during the Sixteenth Century', *PUSH,* vol. 4, no.1, January 2003, pp. 123-33.

———, 'Trade in the Indian Ocean and the Portuguese System of Cartazes', in Malyn Newitt, *The First Portuguese Colonial Empire,* Exeter, 1986, pp. 69-83.

———, 'Indian Shipping and the Maritime Power of the Portuguese in the Seventeenth and Eighteenth Centuries', *Proceedings of the Indian Historical Records Commission,* New Delhi, 1995, pp. 25-34.

———, 'Portuguese Shipbuilding Methods in India in the Sixteenth and Seventeenth Centuries', in B. Arunachalam (ed.), *Essays in Maritime Studies*, vol. II, Mumbai, 2002, pp. 57-61.

———, 'Health-care on Board the Portuguese Ship of the India-Run (Carreira da India) in the Sixteenth Century', *Pondicherry University Journal of Social Sciences and Humanities*, vol. 3, no. 1, pp. 107-17 (January 2002).

———, 'Indian Naval Encounters with the Portuguese: Strength and Weakness', in K.K.N. Kurup (ed.), *India's Naval Traditions (The Role of Kunhali Marakkars*), New Delhi, 1997, pp. 6-25.

———, 'Navigation in the Arabian Sea during the Sixteenth Century', in K.S. Mathew (ed.), *Shipbuilding and Navigation in the Indian Ocean Region* AD *1400-1800,* Delhi, 1997, pp. 26-43.

———, 'Treatises on Portuguese Shipbuilding in India during the Sixteenth and Seventeenth Centuries', *Indica, Platinum Jubilee of the Heras Institute Special Jubilee Volume*, vol. 38, nos. 1 & 2 (March & September 2001), pp. 153-60.

Matos, Luís de, 'O Manuscrito Autógrafo da Ars Nautica de Fernando Oliveira', *Bol. Intern. Bibl. Luso-Brasileiro*, 1, no. 2 (1960).

Matos, Luís, Jorge Rodrigues Semedo de, 'As Navegações Arabes e Portugueses no Oceano Indico, Durante os Séculos XV e XVI', *Mare Liberum*, no. 10, December 1995, pp. 565-82.

McGrail, Sean, 'Portuguese-derived Ship Designs and Methods in Southern India', in *Proceedings of the International Symposium on Archaeology of Medieval and Modern Ships of Iberian-Atlantic Tradition*, Lisbon, 1998, pp. 19-23.

———, 'The Study of Boats with Stitched Planking', in Himanshu Prabha Ray and Jean-François Salles (eds.), *Tradition and Archaeology: Early Maritime Contacts in the Indian Ocean*, New Delhi, 1996, pp. 225-38.

Memorias das Armadas que de Portugal pasaram ha India e esta primeira e ha com que Vasco da Gama partio ao descobrimento dela por mandado de El rei Dom Manuel no segundo anno de sevreinado e no do nacimento do Xrto de 1497.

Mehta, R.N., 'A Note on Some Indian Navigational Terms', *Indica*, XVI, 1979, 208-11.

Mendonça, Henriques Lopes de, *Estudos sobre navios portugueses no séculos xv e xvi,* Lisboa, 1971.

———, 'O Padre Fernado Oliveira e a sua Obra nautica', *Memorias da Academia das Ciencias*, Lisboa, 1898, tomo vii, parte ii.

Mota, A.Teixeira da, 'Méthodes de Navigation et Cartographie , Nautique dans l'Océan Indian avant le xvi sècle', *Studia*, XI, Lisboa, 1963, pp. 49-91.

———, *Instruções Nauticas para os Pilotos da Carreira da India nos começos de século XVII*, Lisboa, 1974.

Naravane, M.S., *Heritage Sites of Maritime Maharashtra, Maritime Activity of the Indians from the Earliest Times*, Maritime History Society Mumbai, Twenty Sixth Annual Seminar 2005, Mumbai, 2001.

Nayeem, Anirudha Ray and K.S. Mathew (eds.), *Studies in the History of the Deccan: Medieval and Modern*, Delhi, 2002, pp. 162-78.

Needham, J., *Science and Civilization in China, Nautical Technology*, vol. 4, part III, Cambridge, 1978.

Ohashi, Yukio, 'The Early History of the Astrolabe in India', *IJHS*, 32, 1997, pp. 199-295.

Oliveira, Fernando, *O Livro da Fabrica das Naus (1580) Facsimile*, Transcription and Translation in English, Academia de Marinha, Lisboa, 1991.

Panikkar, N.K. and T.M.Srinivasan, 'Kappal Sattiram: A Tamil Treatise on Shipbuilding during the 17th Century A.D.', *IJHS*, 7, 1972, pp. 16-26.

Pereira, José Malhão, 'Pandarane, an Important Harbour on the Coast of Malabar', in Lotika Varadarajan (ed.), *Indo-Portuguese Encounters: Journeys in Science, Technology and Culture*, vol. 1, New Delhi, 2006, pp. 388-404.

Pereira, Moacir Soares, *Capitães, Naus e Caravelas da Armada de Cabral*, Lisboa, 1979.

Pillai, T.P. Palaniyappa, ed., *Kappal Sattiram*, Madras, 1950.

Prinsep, J., 'Note on Nautical Instruments of the Arabs', *Journal of the Asiatic Society of Bombay*, vol. 5, December 1836, pp. 784-98 also Garbriel Ferrand, *Introduction l'Astronomie Nautique Arabe*, Paris, 1928, pp. 1-23.

Qaisar, A. Jan, 'From Port to Port: Life on Indian Ships in the Sixteenth and Seventeenth Centuries', in Ashin Das Gupta and M.N. Pearson (ed.), *India and the Indian Ocean 1500-1800*, Calcutta, 1987, pp. 331-49.

———, *The Indian Response to European Technology and Culture*, Delhi, 1982.

Rajamanickam, G. Victor et al., 'Maritime History of South India: Indigenous Traditions of Navigation in Indian Ocean, Andaman and Nicobar Islands', CSIR Project Report, Thanjavur, 1991.

Raju, C.K., 'Kamal or Rapalagai', in Lotika Varadarajan (ed.), *Indo-Portuguese Encounters: Journeys in Science, Technology and Culture*, vol. II, New Delhi, 2006, pp. 483-504.

Ray, Himanshu Prabha and Jean-Francois Salles (eds.), *Tradition and Archaeology: Early Maritime Contacts in the Indian Ocean,* Delhi, 1996.

Reis, A. Estacio dos, *Duas Notas sobre Astrolábios*, Lisboa, 1985.

Rocha Madahil, António Gomes de, *'Um Desconhecido tratado de arte naval portuguesa do século xvii'*, in *Arquivo Historico da Marinha*, vol. 1, no. 4, 1936, pp. 277-93.

Sahai, Baldeo, *Indian Shipping: A Historical Survey*, New Delhi, 1996.

Sarma, Sreenula Rajeswara, 'Indian Astronomical and Time-Measuring Instruments: A Catalogue in Preparation', *IJH,* 1994, pp. 507-28.

———, 'Instrumentation for Astronomy and Navigation in India at the Advent of the Portuguese', in Lotika Varadarajan (ed.), *Indo-Portuguese Encounters: Journeys in Science, Technology and Culture,* vol. II, Delhi, 2006, pp. 505-15.

———, 'The Sources and Authorship of the *Yuktikalpataru*', *Aligarh Journal of Oriental Studies*, 1986, pp. 39-54.

Sastri, Iswara Candra (ed.), *Yuktikalpataru Maharaja-Sri-Bhojaviracitah,* Calcutta, 1917

Soundarapandian, S., *Navai Sattiram* (Tamil), Madras, 1995.

Souza, Viterbo, *Trabalhos Nauticos dos Portugueses no seculo xvi e xvii*, Lisboa 1898, 2 vols.

Stephen, S. Jayaseela, 'Portuguese Nau: A Study of the Cargo Ship in the Indian Ocean during the Sixteenth Century', in K.S. Mathew (ed.), *Shipbuilding and Navigation in the Indian Ocean Region* AD *1400-1800,* Delhi, 1997, pp. 62-81.

Thangal, E.P.A., 'Thangal Calendar in Lakshadweep', in Lotika Varadarajan (ed.), *Indo-Portuguese Encounters: Journeys in Science, Technology and Culture,* vol. II, New Delhi, 2006, pp. 560-73.

Thrower, W.R., *Life at Sea in the Age of Sails*, London, 1972.

Tibbetts, G.R, *The Navigational Theory of the Arabs in the Fifteenth and Sixteenth Centuries,* Coimbra, 1969.

———, *Arab Navigation in the Indian Ocean before the Coming of the Portuguese*, London, 1971.

Tirumalai, R., 'A "Ship Song" of the Late 18th Century in Tamil', in K.S. Mathew (ed.), *Studies in Maritime History*, Pondicherry, 1990, pp. 159-64.

Valente, José Carlos Costa, 'Mentalidade Técnica no Livro Primeiro da Architectura Naval de João Baptista Lavanha (*c.*1600-1620)', *Mare Liberum*, no. 10, Dezembro 1995, pp. 597-606.

Varadarajan, Lotika, 'Indian Rutters', in Lotika Varadarajan (ed.), *Indo-Portuguese Encounters: Journeys in Science, Technology and Culture,* vol. II, New Delhi, 2006, pp. 468-75.

———, 'Indian Boat Building Traditions: The Ethnological Evidence', Topoi, 3/2, 547-68.

———, *The Rahmani of M.P. Kunhikunhi Malmi of Kavaratti: A Sailing Manual of Lakshadweep*, New Delhi, 2004.

Varthema, Ludovico di, *The Itinerary of Ludovico di Varthema of Bologna from 1502 to 1508*, Delhi, 1997.

Waters, David, *The Iberian Bases of the English Art of Navigation in the Sixteenth Century*, Coimbra, 1970.

———, *The Sea or Mariners' Astrolabe*, Coimbra, 1966.

Zheng, Y. Jun, 'As Técnicas de navegação nas armadas de Zheng He e sua contribução Para a ciencia náutica', in *Actas Seminário Ciência Náutica e Tecnicas de Navegação nos seculos XV e XVI*, Macau, 1988, pp. 75-98.

Index